FRONTIERS AND SPACE CONQUEST

FRONTIÈRES ET CONQUÊTE SPATIALE

International Colloquium
Colloque Internationale

FRONTIERS AND SPACE CONQUEST
FRONTIÈRES ET CONQUÊTE SPATIALE

Paris, France — 13-16 January 1987
Paris, France — 13-16 Janvier 1987

Organizing Committee

Jean-Claude Pecker (Chairman)
Roger-Maurice Bonnet
Catherine Cesarsky
Jean-Pierre Faye
Monique Léger-Orine (Executive Secretary)
Jean Schneider (Scientific Secretary)
Isabelle Stengers
Heinz Wismann

Cover picture: With courtesy of the European Space Agency

Frontiers and Space Conquest

The Philosopher's Touchstone

Frontières et Conquête Spatiale

La Philosophie à l'Épreuve

EDITED BY

JEAN SCHNEIDER

AND

MONIQUE LÉGER-ORINE

Centre National de la Recherche Scientifique

Springer-Science+Business Media, B.V.

Library of Congress Cataloging in Publication Data

International Colloquium, Frontiers and Space
Conquest (1987 : Paris, France)
Frontiers and space conquest.

Includes indexes.
1. Outer space--Exploration--Congresses.
I. Schneider, Jean. II. Léger-Orine, Monique.
III. Title. IV. Title: Frontières et conquête spatiale.
TL787.I47 1987 910'.0919 88-9334

ISBN 978-94-010-7845-0 ISBN 978-94-009-2993-7 (eBook)
DOI 10.1007/978-94-009-2993-7

All Rights Reserved
© 1988 by Springer Science+Business Media Dordrecht
Originally published by Kluwer Academic Publishers in 1988
Softcover reprint of the hardcover 1st edition 1988

No part of the material protected by this copyright notice may be reproduced or utilized in any form or by any means, electronic or mechanical including photocopying, recording or by any information storage and retrieval system, without written permission from the copyright owner

CONTENTS

PREFACE

Nous sommes à une époque où les programmes spatiaux connaissent, en dépit de difficultés passagères, une activité croissante. Leur succès dépend de la maîtrise d'un grand nombre de technologies qui de ce fait focalisent l'attention de la plupart des acteurs spatiaux. Toutefois, parallèlement, cette nouvelle aventure nous touche aussi de plus en plus, nous les humains. D'abord, parce que nous sommes les bénéficiaires, sur le plan qualité de la vie (santé, télécommunications, etc...) des techniques spatiales mais aussi parce que de plus en plus les humains vont progressivement étendre leur présence dans l'espace. Pour le moment, le rôle des astronautes est essentiellement d'être des "travailleurs de l'espace" plus efficaces que les robots les plus perfectionnés. Mais inévitablement l'aspect proprement humain de ces hommes et de ces femmes ressurgit : d'abord pour eux-mêmes, mais aussi pour nous car, par le jeu des processus d'identification, ils emportent avec eux une parcelle d'humanité et donc de nous-même.

Il n'est donc pas inutile de réfléchir à ces questions et le but du présent colloque était de le faire de manière approfondie en faisant appel à des professionnels, c'est-à-dire les philosophes et les penseurs des sciences humaines.

"La philosophie à l'épreuve", tel est le sous-titre de ce colloque. Mais de quelle philosophie s'agit-il ? Ce terme peut renvoyer à deux conceptions très différentes : d'une part la philosophie naturelle, d'autre part la philosophie critique. On peut dire que les deux se distinguent par le fait que la première attribue du <u>sens</u> ou de la <u>signification</u> aux phénomènes naturels alors que la seconde réserve cette dimension à la sphère de l'humain. Pour simplifier, la première conception est celle que l'on rencontre principalement en URSS et aux USA, la seconde étant prédominante en Europe. Selon le premier point de vue, l'humain est un prolongement naturel de processus physiques naturels. Selon le second, il faut présupposer des catégories conceptuelles antérieures à tout objet naturel auquel elles s'appliquent, ici la conquête spatiale ; elles sont de ce fait à l'abri de toute évolution qui serait imposée par cette conquête.

Le présent colloque a été l'occasion d'une confrontation entre ces deux points de vue. Il avait pour intention de départ, suggérée par son sous-titre, de poser la question "quels remaniements la conquête spatiale est-elle susceptible d'opérer sur notre univers mental ?" Mais il ne s'agissait pas d'enfermer a priori les intervenants dans cette formulation. Elle constituait plutôt un horizon, les intervenants étant libres de leur propre problématique.

Pour terminer, il m'est agréable de rendre justice à l'opiniatreté de J.P. Faye dans son projet d'Université Philosophique Européenne qui, au cours de l'histoire mouvementée de la gestation de ce colloque, a su lui donner l'impulsion nécessaire à un moment décisif.

Jean Schneider

FRONTIERS AND THE CONQUEST OF SPACE
INTRODUCTION

Reimar Lüst
Director General
European Space Agency
8-10, rue Mario-Nikis, 75738 Paris Cedex 15, France

Some people might be surprised or at least might find it remarkable to receive a joint invitation by two institutions which have activities in completely different fields, philosophy on the one hand, and technology on the other. Indeed we cannot deny that our western culture has split up into two diametrical parts. This fact has already been stressed upon by C.P. Snow in 1959 in his famous Reith-Lecture dealing with the two cultures and the scientific revolution. While one of these cultures is carried by the art, literature and humanities group, the other is with the natural scientists. Each of these groups has developed within its own culture common conceptions, apprehensions, behaviour and measures, but these two cultures are separated by a deep cleft of nonunderstanding each other.

But it is increasingly recognized in our time that we are making great strong efforts to overcome this gap. The only way I see to achieve this is to speak to each other and to explain to one another the ongoing developments and the existing problems. Therefore I was very pleased when I heard about the idea of this colloquium and I am very happy that it will take place during this week here in Paris. On behalf of the European Space Agency I should like to welcome most heartily everybody at this colloquim.

I should like to make some remarks for the introduction of this colloquium "Frontiers and the Conquest of Space" as some inputs for the dialogue between the two groups present here at this colloquium.

1. What is basically new for mankind here on Earth since the first man-made satellites circling around the Earth and the first space probes penetrated into our planetary system? Let me mention three points in this aspect:

   - for the first time in history, man is able to surpass the gravitational atmosphere of the Earth and really to leave the Earth. In this way he can express in a new and unique way the finiteness of the Earth and the infinity of the Universe;

J. Schneider and M. Léger-Orine (eds.), Frontiers and Space Conquest, 1–2.
© 1988 by Kluwer Academic Publishers.

- for the first time man can observe the Earth regularly from outside as well as we can see the Universe without the hindering influence of the Earth's atmosphere;

- by going into space we are able to overcome natural boundaries and we are able to communicate in a global way.

These three important steps which have been achieved by developing and using space technologies can in principle be done without sending man into space, but have important consequences for us living here on Earth.

2. This brings me to the second aspect, namely that we as human beings have the possibility to leave the Earth, to stay in space in a satellite or space station circling around the Earth, or to go to other places in our planetary system as has already been done by going to the Moon.

   We are discussing here in Europe just now whether we should embark on the manned spaceflight programme, or whether we should leave the occupation of space by man entirely to the two big space powers.

3. The third aspect which I should like to mention in introducing this colloquium is the fact that it is organized by two European institutions, the European Philosophical University and the European Space Agency. The bond between the European nations is our common culture and we have to make sure that we can also defend this for the future. In doing so we not only have to remember this, but to develop this further with our own ideas and with our own spirit.

The successful work of the European Space Agency is a very good demonstration that we in Europe can work together and that in this way we can build up Europe. I am convinced that the European Philosophical University can also make very fundamental contributions to the process here in Europe of finding a better way of working together and of pursuing our way of life.

I would hope that this colloquium will also be instrumental in helping to find reliable bridges between the two cultures, the culture of literature and the culture of scientists and technologists.

POURQUOI CE COLLOQUE ?

Jean-Claude Pecker
Professeur au Collège de France
11, place Marcelin Berthelot 75007 Paris, France

Quand on parle de conquête spatiale, il est automatique de se référer à Christophe Colomb. A Colomb, et à la foule de conquistadores de toute espèce qui se sont jetés, sous le vent de ses bateaux, comme une nuée de sauterelles sur le Nouveau Monde. En peu d'années, ils en firent l'un des plus lamentables déserts de vieilles pierres qui soit, après un génocide quasiment radical, avant d'y installer une colonisation galopante qui est, en effet, devenue notre Nouveau Monde.

A mon avis, il ne faut pas trop exploiter ce précédent comme base de comparaison valable entre révolution qui s'amorce sous nos yeux, et nous.

Il n'y a pas, à vrai dire, la moindre analogie entre ces deux périodes de notre histoire ; il faut se méfier de l'abus de références historiques ! Paul Valéry avait bien raison d'affirmer que l'histoire donnant des exemples de tout, elle ne peut donc nous apprendre rigoureusement rien...

Que de différences en effet, en y regardant de plus près, entre l'Armada d'Isabelle la Catholique et celle des grandes agences spatiales.

Lors du départ de Colomb, ce qui était nouveau, c'était seulement le cap sur lequel naviguait le Gênois, l'Ouest. La mer, tous les marins de la petite flotte la connaissaient déjà, avec ses tempêtes et la monotonie de ses calmes plats. Le moyen, des caravelles sans innovations remarquables. Le but, le Cathay, la Chine, visitée depuis les voyages vers l'Est de Marco Polo ou des explorateurs arabes. La seule surprise fut en somme la découverte fortuite de nouvelles terres habitées.

Pourtant, la conquête de l'Amérique puis, aux XVIème et XVIIème siècles, la multiplication des explorations, des conversions, des liens maritimes et commerciaux entre des contrées à portée de voile les unes des autres, tout cela annonçait et préparait les temps modernes.

La conquête spatiale d'aujourd'hui annonce et prépare nos temps futurs, certes ; mais elle est d'une toute autre nature.

A l'exploration d'une terre finie dans sa sphéricité définitive, succède l'acquisition d'une troisième dimension du grand voyage ; c'est une exploration <u>en profondeur</u> qui ne se heurtera pas de sitôt à la sphéricité très lointaine de l'univers peut-être fini des relativistes.

J. Schneider and M. Léger-Orine (eds.), Frontiers and Space Conquest, 3–8.
© 1988 by Kluwer Academic Publishers.

A l'exploration à pied, à cheval, en bateau ou en voiture, voire en avion, succède une exploration en fusée-sonde ; on s'écarte franchement de la Terre. Toute référence à la surface terrestre, plan horizontal ou machine ronde, est supprimée. Si l'on veut ici formuler une analogie, c'est avec la petite enfance de l'homme, voire avec sa naissance, qu'il faut désormais comparer. Faut-il psychanalyser la recherche spatiale, nouvelle naissance, -Re-naissance ? C'est Tsiolkowsky, je crois, qui disait il y a plus d'un siècle que "*la Terre est le berceau de l'humanité ; mais* [que] *nul ne peut éternellement rester au berceau*". La métaphore est insistante : elle nous interroge.

N'aurions-nous vecu qu'une préhistoire, ou qu'une gestation, avant l'avènement de la conquête spatiale ?

Au bout du voyage spatial, nulle découverte fortuite d'une terre inconnue. Nous visiterons un monde déjà exploré par le télescope et des engins automatiques d'exploration. Peu de grandes surprises à attendre du paysage ou des habitants... En revanche, ce sera l'exploration, dans ses moindres détails, d'un milieu autre, d'un sol inconnu, d'une chimie, d'une botanique, d'une zoologie nouvelles ; puis ce sera sans doute l'exploitation de ce nouveau "nouveau monde", mais si tardive dans la nouvelle conquête qu'elle ne saurait affecter l'idée qu'aujourd'hui nous pouvons nous faire de cet abandon de la Terre-mère.

Exploration minutieusement préparée, mais conditions quotidiennes de voyage et de vie radicalement nouvelles.

Sans doute, ne nous en apercevrons-nous que peu. Sans doute, nos enfants eux-mêmes ne verront-ils là qu'un nouveau spectacle, ou un nouveau jeu électronique. Sans doute, l'économie mondiale n'en sera-t-elle pas immédiatement affectée.

Mais avant même que le voyage ait dépassé la proche banlieue, je veux dire la Lune, l'importance de la conquête spatiale s'inscrit déjà dans nos politiques militaires (je me réfère bien sûr à l'initiative de défense stratégique) ; nous la voyons, dans une certaine mesure, dans nos littératures et peut-être dans certaines attitudes quotidiennes de notre vie politique : ne nous dit-on pas parfois que la compétition pour la conquête de l'espace contraindra les grandes nations spatiales à s'unir ? Est-elle le prélude à un gouvernement mondial ? A une entente au moins ? On voudrait le croire ; des signes semblent l'amorcer.

Seulement, un regard sur l'échiquier mondial ne peut pas ne pas jeter l'inquiétude en nos coeurs, peut-être trop tôt à l'écoute d'un XXIème siècle spatial, celui que préparent activement nos ingénieurs. Les possibilités militaires des grandes puissances sont telles qu'une guerre dont l'enjeu serait l'espace n'est guère plus concevable que la guerre contre les éventuels extraterrestres (pas plus, en termes anciens, la guerre entre colonisateurs que la guerre de colonisation). Mais devons-nous pour autant espérer bientôt la fin des conflits terrestres ? Personnes n'en croit malheureusement rien. Les guerres locales ne changeront guère de nature. Elles conserveront leur image de l'horreur quotidienne.

Il est bien certain aussi que, dans les campagnes de l'Inde ou sur les plateaux péruviens, on ne s'apercevra pas de sitôt que quelque chose est en train de changer l'humanité en profondeur. Le satellite ECHO, par exemple, bien visible naguère en sa promenade, n'a apporté qu'un astre de plus, bien vite dégonflé et oublié.

La révolution spatiale, la conquête et le recul permanent des frontières n'entraîneront sans doute aucune mutation immédiate du quotidien, donc de l'humain. On pourrait alors penser que notre réunion ici et aujourd'hui est prématurée.

Pas si sûr !

Car c'est, je crois, de ses conséquences secondaires et non des implications directes de cette conquête que pourra venir l'imprégnation. Nos ingénieurs, nos stylistes, nos artistes, nos biologistes, nos médecins vont, <u>à l'occasion</u> du voyage spatial, inventer des solutions à de nouveaux problèmes. Ces solutions donneront aux Terriens restés à Terre des outils, un style de vie même, qui seront vite profondément marqués et de façon insolite, par ces technologies très pointues. On imagine a priori des exemples nombreux, et des expériences ayant déjà eu lieu seront sans doute évoquées ici-même. De nouveaux matérieux permettront -pourquoi pas ?- la réalisation de quelque bicyclette ultra-légère qui battront tous les records : les voitures mises au point pour rouler sur la Lune serviront au Sahel : sa mise en culture, grâce à des engrais fabriqués sur Terre, se développera peut-être à l'aide d'une chimie empruntée au milieu naturel vénusien. D'étranges cuisines s'acclimateront sur Terre ; ce sera une curiosité, et l'on imagine aisément le développement d'un certain snobisme spatial : ici, on servira un repas concentré d'astronaute ; ailleurs, ce sera plutôt le lichen du Chef à la martienne, qui triomphera ; plus prosaïquement, certaines idées de conservation, ou de présentation, d'aliments sous faible volume seront appliquées à la consommation courante et l'on trouvera assez vite les produits nouveaux, aux étalages des supermarchés -pardon, je veux dire des "space-shops" ! Des tissus créés pour les cosmonautes, seront utilisés en haute couture ; d'autres, moins onéreux, bénéficieront sans aucun doute des progrès accomplis. Notre cadre de vie sera modifié lui aussi : car peut-on penser que la réflexion des stylistes en mobilier, qui se sera largement exercée dans les navettes, pourrait ne pas tenter d'imposer ses marques nouvelles à notre habitat ? En fait, ce sont les techniques des transports, des industries de l'agriculture, de la chirurgie, de la pharmacopée, et de tant d'autres qui seront très vite bouleversées.

La façon de penser des astronomes, des géophysiciens ou des géologues, sera différente, comme le sera également celle des physiciens, des chimistes, des biologistes et de tous les spécialistes des techniques avancées. Cela est l'évidence même ; et bien que ce ne soit pas l'objet de notre colloque, je ne peux pas ne pas évoquer cet aspect du progrès des techniques spatiales, au moins dans mon domaine, l'astronomie.

C'est d'abord une exploration in situ des planètes, des milieux interplanétaires, et donc la compréhension des processus essentiels de la formation des systèmes planétaires en général et, dans ces systèmes, de l'apparition physicochimique de la vie. Ces recherches modifient profondément la situation épistémologique de l'astronomie. En effet, de science d'observation passive, l'astronomie devient expérimentale et active.

L'astronomie spatiale implique en second lieu la possibilité de disposer facilement d'observatoires très complets hors de notre atmosphère trouble et diffusante, hors même du plan de l'orbite terrestre, donc loin des poussières interplanétaires diffusantes et perturbatrices qui encombrent ces régions de l'espace. En fin de compte, c'est la possibilité de gagner de façon considérable en sensibilité et en pouvoir de résolution. Entre autres problèmes astrophysiques dont l'aspect sera bouleversé grâce à ce gain de sensibilité : la cosmologie, servie par une pénétration aujourd'hui impossible des profondeurs de l'univers et une remontée dans le temps vers de nouveaux témoins, extrêmement anciens, de son évolution. Il est difficile de prévoir la nature de ces observations lointaines qui multiplieront vite par dix, cent, ou plus, le volume de l'univers exploré. Les cosmologistes font à cet égard des prévisions différentes : tous attendent de l'espace des clefs essentielles de leur description du "tout cosmique". Leur façon même de penser en sera profondément affectée.

Mais celle des philosophes, et tout simplement celle des hommes, des Terriens, sera-t-elle aussi différente ?

Pour ce qui est des philosophes, qu'ils veuillent bien me pardonner si je ne leur dis pas à l'avance de quoi ils vont nous entretenir cette semaine !

Pour ce qui est de l'homme ou de la femme de la rue, on peut se livrer à quelques conjectures. Le mode de vie, même considérablement transmuté, n'entraîne pas nécessairement une nouvelle façon de se poser les problèmes ontologiques : essence, existence, vie, et mort, pourquoi ? Vers quoi ? Comment ? Et la suite... Pourtant, nous disposons déjà d'une certaine expérience, sans valeur statistique je l'admets, mais peut-être significative. Rappelons-nous par exemple la grande peur martienne des New-Yorkais, déclenchée par une émission radiodiffusée d'Orson Welles dans les années d'avant-guerre. Evoquons ce vieil homme du fond de l'Arkansas, je suppose, qui, après avoir vu Armstrong poser le pied sur la Lune, restait convaincu qu'il s'agissait d'un trucage, car Dieu avait interdit aux hommes de quitter la Terre. Certes, les enfants d'aujourd'hui ont bien pris le pli ; la guerre de l'espace, tout comme sa conquête, sont objets de jeux, et non pas de mystère, ni de mystification ; et d'avoir accepté Armstrong, perçu d'abord comme le parangon de l'impossible héroïsme, pousse maintenant nos enfants à croire en Goldorak, véritable chevalier, quand ce n'est pas en quelque autre homme-dieu, Moon, Khaddafi ou Rambo, en quelque magie transmutée en science. Un moderne Prométhée ? Bien plus sans doute que le progrès technique, c'est cette composante de l'imprévisible, de l'extranormal, qui le rejettera vers d'étranges affabulations, expériences de pensée (en somme) plutôt que, comme naguère, revanche envers quelque dénuement intellectuel.

Si bien que l'espace apparaît aussi bien comme un recours que comme une menace. Ainsi, pour certains, la conquête spatiale est-elle l'alternative à l'apocalypse nucléaire, voire à l'explosion démographique. Je citerai simplement Arthur Koestler qui conclut ses Sleepwalkers ainsi : "*dans un avenir prévisible, l'homme se détruira, ou partira pour les astres*,"... prévision, d'une nécessité qui s'imposera à l'espace humaine. Mais l'ambigu Koestler achève en prévenant le lecteur contre les dangers "*de l'hybris de la science, ou plutôt de la conception philosophique que l'on fonde sur elle*". Une science, un pouvoir technologique, qui seraient pour Koestler, et c'est la dernière phase de ce livre qui fit date, un "*nouveau Baal qui règne sur le vide moral du haut de son cerveau électronique*". Et l'angoisse, qui nous contraindrait à quitter une Terre trop ou mal peuplée, ne nous étreint-elle pas aussi à l'idée même du départ ? Comment ne pas évoquer les accidents qui ont endeuillé les premières tentatives de vol spatial, et le dernier, tout récent, encore présent en nos coeurs, et qui a contribué à de dramatiques remises en question, la terrible explosion de Challenger ?

A nos philosophes, à nos psychologues, à nos sociologues de nous dire maintenant ce qu'il faut penser de tout cela. Je m'en veux un peu d'avoir retardé leurs remarques. Elles correspondent pour nous à un véritable besoin : nous savons à peu près quand nous enverrons quoi et qui dans les espaces interplanétaires, voire intersidéraux. Mais nous n'avons que des idées vagues des conséquences humaines de ces lancements divers, -autant de bouteilles à la mer spatiale ! Et nous posons la question. Dans un tel débat, les protagonistes se trouvent un peu au sommet ambigu d'une moderne Babel. Les uns analysent propergols, gravité, plasmas et microélectronique, les autres discutent de comportements, d'angoisse, d'être et peut-être de néant.

Pendant plusieurs mois, dans cette maison même, un séminaire préparatoire a permis de faire converger, dans une certaine mesure, vocabulaires et points de vue, au moins à l'échelle d'une petite communauté. Le présent colloque est un effort d'extension de ce dialogue à une communauté plus large, la communauté européenne, qui aborde d'un front commun les efforts de conquête spatiale du présent et de l'avenir, comme l'a souligné le Professeur Reimar Lüst.

Mais ce colloque est aussi un effort de création, un acte -pourrait-on dire- chargé d'implications sociales et politiques, destiné à forcer en quelque sorte les communautés scientifiques et philosophiques à sortir du cadre de leurs préoccupations traditionnelles et parfois... ronronnantes.

Que les scientifiques s'intéressent aux aspects humains de la science et à la philosophie ! Que les philosophes consacrent aux problèmes de leurs temps l'intérêt qu'ils méritent ! Notre ambition est grande, et ce colloque n'est qu'un premier pas.

L'idée du Colloque remonte à une proposition faite par M. Delamare, et par M. Hervieu, alors tous deux au Ministère de l'Industrie et de la Recherche. Cette proposition fut étudiée par une instance du Collège International de Philosophie. Un comité fut mis sur pieds. Pour abréger, je voudrais maintenant exprimer ma reconnaissance personnelle aux initiateurs déjà nommés, à ceux qui ont participé aux premières réunions, comme à ceux qui, après élagage de certains thèmes, ont organisé d'abord notre séminaire interdisciplinaire, puis ce Colloque, et notamment les membres de son Comité d'Organisation, Roger Bonnet, Catherine Cesarsky, Jean-Pierre Faye, Isabelle Stengers, Heinz Wismann, et bien sûr celui qui fut l'âme du séminaire, et qui est l'âme du Colloque, Jean Schneider, comme celle qui en est le bras séculier, Monique Orine, ainsi que tous ceux et toutes celles qui ont contribué à l'animation du séminaire et aux travaux d'organisation.

La responsabilité de l'organisation et du financement a été assumée par l'Agence Spatiale Européenne et par l'Université Philosophique Européenne. Je tiens à demander à Reimar Lüst et à Jean-Pierre Faye d'exprimer notre gratitude à leurs administrations respectives ; à eux-mêmes, je tiens à dire à quel point leur appui permanent nous fut précieux, comme celui d'Yves Laporte, Administrateur du Collège de France, d'André Berroir, Directeur Scientifique du CNRS, et de Tor Bloch, Directeur du Centre de Calcul Vectoriel pour la Recherche, qui a mis Monique Orine à notre disposition. Enfin, je voudrais dire combien nécessaires furent les encouragements moraux et matériels qui nous ont été prodigués par les ministres successifs responsables de la Recherche, Messieurs Laurent Fabius, puis Hubert Curien, Ancien Président du CNES, qui a bien voulu accepter, à ce titre de tirer les conclusions de nos débats, et enfin Alain Devaquet. C'est dans le bel amphithéâtre Henri Poincaré de ce Ministère que nous pouvons nous réunir aujourd'hui.

Et bien sûr, je remercie tous ceux qui sont ici, venus parfois de bien loin, de toutes les richesses dont ils répondront à notre interrogation collective.

Alors, "pourquoi ce Colloque" ? J'ai répondu de façon très partielle : d'abord, je viens de le dire, parce que la machine a été mise en marche par certains. Bien sûr, plus fondamentalement, parce que l'adaptation de notre monde aux nouvelles possibilités des techniques spatiales se pose, je l'ai dit tout à l'heure, en des termes très concrets. Et aussi certainement pour bien d'autres raisons plus profondes, que j'ai évoquées, mais que seule pourra dégager la conclusion de nos débats.

Sans plus attendre, nous allons donc maintenant entrer dans le vif du sujet...

AFTER ICARUS

André Lebeau
Professor at the Conservatoire des Arts et Métiers
(Chair of Space Technology and Space Programmes)
292, rue Saint-Martin 75141 Paris Cedex 03, France

By way of introduction, I am going to attempt to give an overall view of the way in which space technology interacts with the development of society, most notably in the developed countries where this technology is being created. I might have entitled my address "The logic of space development". There is no intrinsic logic to the development of a technology; logic becomes apparent only if one examines the whole constituted by a particular technology and its socio-economic framework. And that is just what I am proposing to do. I shall be seeking throughout to disregard circumstantial factors in order to concentrate on essentials. I will also avoid getting involved in an analysis of national strategies, since that would lead me way beyond the limits I have set myself. I will be using a historical form of presentation, the method best suited for demonstrating the logic of space development. I will start by going back to the beginnings.

Space technology, unlike some other technologies, was born out of a conjunction of factors, and not in response to an express need -as could be said to have been the case with the steam engine, for example. Nor was it the result of a scientific breakthrough opening up a new area of technology. The conceptual apparatus needed for conquering space was already there at the end of the 17th century. This we owe, as everyone is aware, to Newton who had, moreover, realised that his work was leading towards the possibility of creating an artificial celestial body. If we look at the work of the "precursors" in this area, those who applied themselves -from the closing years of the 19th century until the end of the thirties, just before the outbreak of the second World War- to the practical problems of conquering Space, the most striking feature is the enormous discrepancy between the resources available to them and their technical objectives, which may be summed up in two figures: attainment of an altitude of around 300 km and a horizontal speed at that altitude of 8 km per second. Goddard's rocket in 1926, exactly 60 years ago, provides a graphic illustration of how far we have come within a single human lifetime, and a mere look at that rocket suffices to give some idea of the scale of the discrepancy in Goddard's own time between the resources available and the goals people were setting themselves. Those goals were extremely ambitious: people were already dreaming of occupying and even colonising Space and the celestial bodies, objectives that were manifestly too remote to stimulate any

J. Schneider and M. Léger-Orine (eds.), Frontiers and Space Conquest, 9–17.
© 1988 by Kluwer Academic Publishers.

significant response on the part of governments and still less from the private sector. Such targets are, in a certain way, beyond the range of the forces that drive markets and governments.

Space was born out of the conjuntion of military ballistics and the fundamental reguirements of space technology: it was this meeting that made it possible to emerge from the impasse that had been reached. The combination of ballistic missile and thermonuclear charge -which has dominated military strategy for three decades- provided the key to Space, with a minimum of additional effort. These highly particular origins have had two consequences. In the first place, since space technology came into being almost by chance, without being created to meet an existing need, it was initially something of a "solution looking for a problem", to borrow the title of an article that appeared some years ago in Le Monde: "Laser, a solution looking for a problem". At the time when the article was published (during the 1960s), people knew how to produce lasers but there were as yet few applications for the technology. Everyone knows, without there being any need to dwell on the point, that the position has altered radically in the meantime, and the same applies to space technology. I would like to underline this point by emphasising that, contrary to a fairly widely-held view, military interests were not in fact particularly influential in the development of space technology at the outset. A report [1] published in 1947, under the title "Preliminary design for an Earth-orbiting spacecraft", by the Rand Corporation or, more exactly, the nucleus of what was later to become the Rand Corporation, working at the time for the United States Air Force, reached a negative conclusion, in terms of cost-effectiveness, as to the military value of such an undertaking. The report also included the following interesting assessment: "*If the United States were to build a satellite, the human imagination would be set on fire, with likely effects worldwide, comparable to those following the explosion of the first atomic bomb*". I shall come back to this comment.

The second consequence of these mixed military/chance beginnings is that the role of the state in space technology has been perpetuated -I am referring of course to the West and not to the Communist world where there is, in practical terms, no alternative to the role of the state- and this for two reasons: the first, long-standing one relates to the secrecy attached and still attaching to military ballistics technology. It has taken the gradual divergence of military ballistics and space launchers to reduce the burden of such secrecy. The second factor is the absence of immediate commercial objectives, which has meant that states have been the only driving force behind space technology development. The military origins are still very evident in the range of launchers in service in the United States. The Delta, Atlas and Titan, three of the most heavily used launchers, mark three stages in the build-up of United States nuclear ballistic capability. The Delta is based on a single stage, which was the first medium-range missile: the Thor; the Atlas is based on the first intercontinental ballistic missile, and the Titan corresponds to the second generation of intercontinental missiles.

The historical development of space technology may be broken down into three successive phases: we are now at the start of the third. Each phase is distinguished by the driving forces behind its technological development.

The impetus for the initial phase -which lasted until around the end of the 1960s, although there is obviously no fixed boundary- came from three sources. First, the interest shown by certain scientific disciplines. This interest was, in fact, apparent from the outset and played a not insignificant role in the setting up of the first launches, on the American side in particular. Those first launches took place against the background of the International Geophysical Year; from the very beginning of the US programme, the scientific interest, a permanent feature of current space programmes, has been instrumental in the development of a branch of activity represented by the Explorer, Pioneer and Mariner programmes, that have performed pioneering work in extending knowledge of the Earth's environment and the solar system.

The second driving force is the military interest. As I have already said, this was initially minimal. However, that did not last long. If one looks specifically at the US programme, about which more is known, it is evident that little time was lost there in recognising the major potential applications for space technology. The application which led directly to the start-up of a US military programme in this area concerns the problem of locating Soviet cities. A military problem, intimately linked to the strategic doctrine of "mutual assured destruction", which has dominated international relations ever since, based, inevitably, on the thermonuclear ballistic missile. To implement this doctrine of mutual assured destruction requires knowledge of the exact whereabouts of the targets. The targets in this instance are centres of population, i.e. cities, and while the Soviets were well-informed as to the location of US cities, the Americans did not have sufficiently accurate knowledge of the position of Soviet cities. If their destructive capacity was to be credible, the Americans needed to be able to pinpoint such centres. The programme conceived for this purpose was the Discoverer programme, based on the principle of obtaining images of Soviet territory which could be related with sufficient accuracy to the satellite's geodesic position to satisfy nuclear missile targeting requirements.

The third and most important driving force behind the space programme's initial phase: the superpowers' search for prestige, an element which, as we have seen, had been accurately predicted in the Rand Corporation's report. Political circles, in the West at least, had been less clear-sighted. The impact on American public opinion of the first Soviet launch triggered off a phase that may be described as a period of ideological confrontation through the intermediary of space achievements, culminating in the Moon race and the symbol of the end-of-the-decade, the cornerstone of Kennedy's famous speech on Space.

This confrontation brought a very rapid expansion in space budgets. The dramatic decline which set in in 1968, against the background of a transformed world economy, marked the end of the first phase. It was the result of the coming together of a number of factors, some of which were foreseeable, others less so.

First, an absence of motivation on the American side. The Soviets had decided, for reasons that are difficult to analyse, to drop out of the Moon race, and the United States realised that they were running on their own. Secondly, there was a lack of economic, technological or scientific rationale for involving astronauts in exploration of the planets and other bodies. Mention should also be made -and it is a point worth noting- of the highly prejudicial effect on the conception of the Apollo programme of choosing the end-of-the-decade symbol as a target for the undertaking in which the American nation was engaged. This resulted in the adoption for conquering the Moon of a technological system that could not be re-used for conquering other celestial bodies, such as Mars. For more ambitious undertakings it was therefore necessary to start again from zero. A very clear example of the harm that can be caused by sometimes irrational political decisions when applying in the technological sphere.

The second phase in the development of space technology -a phase that begins is the latter half of the sixties, taking a firm hold in the seventies- is characterised by space applications, the use of space to provide civilian as well as military services. There is a certain unity about these applications; they are <u>all</u> applications of an "informational" kind -they relate exclusively to the handling of information.

What is the underlying logic of these applications? Relatively simple, when reduced to its essentials. The availability of basic space technology offers three immediate opportunities.

The first involves becoming free of the shield constituted by the atmosphere and moving beyond it to observe the universe. The second consists in the ability to move freely about the solar system, taking measurements and observations. And the third opportunity -which sounds like a truism- is that of being able to be at a distance from the Earth, which behind makes it possible to observe it as a celestial body and to use a remote relay to collect data transmitted from the Earth's surface, either naturally or as the deliberate result of human action.

The first two cases have served solely to extend the frontiers of knowledge, to obtain information on the outer universe. The third serves as the basis for all current space applications, civilian or military, which I describe as traditional, even though they are very recent.

As I have already pointed out, these applications share an informational role. Whether it is a matter of telecommunications or Earth observation for civilian purpose, e.g. meteorology or <u>remote sensing</u>, to determine what is happening on the Earth's surface, or observation for military purposes, the role of the satellite remains the same: it gathers information from the Earth's surface and redirects it to the surface again, once it has been processed, transformed or compressed as necessary. These applications fall into two broad categories: either the signal is transmitted naturally -or in any event not deliberately- to the satellite, the whole area of Earth observation, or the signal is transmitted deliberately to the satellite, which constitutes a relay designed to reflect it back to Earth at a location out of direct sight of the point of transmission, the telecommunications category.

I do not intend to go into detailed analysis of the development of these applications, but simply to note some of their most important features. The civilian sector has been dominated by the breakthrough in space telecommunications, progressing extremely rapidly and extending into new fields. Intercontinental telecommunications were the first to be affected: before the advent of space technology and given the state of the Earth-bound technology in the early 'sixties, there had been no other method for transmitting telephone calls or television pictures over the oceans.

The very rapid growth in the volume of intercontinental space telecommunications reflects less an increase in demand -the demand predated the emergence of this technology- and more the increased capacity of space technology to meet the demand.

A similar trend can be observed in the military domain, though proceeding at a much slower pace. Two successive phases can be distinguished: after the initial phase, which I have already mentioned and the principal objective of which was to locate Soviet cities, came a strategic observation phase, the results of which can be seen as having been generally very beneficial. This latter phase enabled each of the superpowers to gain a more accurate impression of the potential strength of its adversary and so of the threat posed. The overall result has been to make strategic arms limitation agreements (SALT) possible -these naturally require that each of the partners have a monitoring capability- so permitting a slowdown in the strategic arms race. At the end of the second phase, we are witnessing another trend: the increasingly widespread use of military satellites, no longer confined to strategic areas and becoming ever more involved in tactical problems, in weapons systems and the actual waging of battle. This has brought a very rapid growth in military space activities. In 1982, the military share of the United States federal budget devoted to Space outstripped the non-military share, and the grap has gone on widening ever since.

We stand today on the threshold of what might be thought of as a third phase, still a virtually closed book to us, although we can attempt to define its main characteristics.

The first thing to note is the changing concept of space transportation, heralded by the arrival of the United States Shuttle. We are seeing a system of space transportation which, unlike the systems which served as vehicles for the development of traditional activities -automatic systems, without crew, where the launcher is burnt up during use- will make the presence of human operators working in Space a real possibility. This is a path on which the United

States, Europe, the Soviet Union and Japan have all set out, making the trend towards use of crewed launchers a general phenomenon.

There should be no confusion, and this is a point I wish to emphasise, between the specific weaknesses of the US Shuttle, weaknesses that are inextricably linked to the choices made when the system was designed, and the characteristics particular to crewed transportation systems. Crewed transportation systems have their own characteristics compared with the automatic systems, use of which has predominated until now. These features do not coincide exactly with those of the Shuttle, which has a number of weaknesses peculiar to itself.

Another element, which would seem to be of fundamental importance, is the discrepancy between this new -government-funded- orientation of space technology, and the needs of traditional space applications.

The discrepancy is first and foremost a qualitative one: traditional applications do not require a new transportation system. For their purposes, conventional launchers of increasing refinement will remain satisfactory for the foreseeable future. It might even be maintained that artificial use of the Space Shuttle on a commercially competitive basis led, in that particular instance, to the catastrophe, a catastrophe that was unquestionably linked to specific flaws in the United States Shuttle, but which also bore witness to the unsuitability of crewed launchers to meet the requirements of traditional applications.

At the same time, the discrepancy is quantitative. It is very unlikely that information-related applications will ever be able to counterbalance the operating costs of crewed systems, to say nothing of the costs of developing these new systems. We are being asked to lay odds on the future and to answer a question: can we imagine fresh areas for space applications opening up for which the new transportation systems would be essential? This brings us to the third aspect of this phase in space development, the prospects for new applications.

We need here to draw a distinction between medium-term and long-term prospects. I shall start with the latter.

The long-term prospects are based on the general idea that Space constitutes a kind of deposit. Of energy, the solar energy that crosses space at the level of the Earth orbit, at 1.4 kW per m^2 perpendicular to the direction of the Sun. Of materials, on the asteroids and the Moon. And a source of physical space as such, which, it is plain to see, is beginning to be in very short supply for certain activities on the Earth's surface.

Over against this notion of Space as a deposit, we need to set the fact that the technological system currently operating in developed societies is confined to the Earth's biosphere, which fact, in terms of resources -e.g. exploitation of fossil fuels- and in terms of damage done to the environment, puts its future in jeopardy. Many works have been published on this theme, some of them wildly exaggerated in their assessment of the future -I need only refer to the Club of Rome's publications, somewhat devalued by being over-dramatic. Nonetheless, the technological system as it is evolving in developed societies cannot be sustained within the biosphere for an indefinite period -I am talking in terms of centuries, consistent with a future on the same scale as humankind's historical past. And this is without even contemplating the whole of humanity enjoying the same level of resources as the developed countries. Thus, the crucial question is whether expansion of our technological system into the solar system, outside the Earth's biosphere, might offer a solution.

You will notice that I am not talking here about colonising Space but about its occupation by human operators, involved mainly in production work. The very presence of such human operators is a problem requiring consideration. It may be wondered whether this expansion of the technological system into the solar system could not be carried out using just automata and robots. You will no doubt be returning to this issue; what can be said very briefly is that there are no obstacles in theory to a technological system that maintains and replicates itself. So much has been apparent since von Neumann, though it may also be stated that opinion is uniting around the view that full-scale robotisation is an improbable scenario for extending the technological system into the solar system and that the expertise of the human operator in Space will in fact be needed. Which is precisely the approach underlying the new space transportation systems.

As things stand at the moment, the main problem is one of progression. Is there a series of stages offering a plausible progression from the current state of development of space technology to a state where a large proportion of production work would be carried out by elements of the technological system located outside the biosphere? Are there intermediate stages between the point where we are now and that distant goal, which cannot be achieved all in one go? This brings me to the medium-term prospects.

Although important, I intend to leave to one side the use of the new space transportation systems and human operators for purely scientific purposes, such as the advancement of fundamental research: these activities, though valuable and altogether respectable, will never be conducted on a scale sufficient to supply an adequate base for such an undertaking. We may identify two possible areas of expansion, very different from each other. The firs relates to the production of materials under microgravity conditions and the second to the development and possible deployment of antiballistic weaponry in Outer Space.

Both of these activities require the presence of human operators, for reasons on which it would be easy to elaborate. They share an element of uncertainty. Regarding the production of materials under microgravity conditions or, more exactly, the economic viability of manufacturing certain materials under microgravity conditions in factories operating in orbit, the uncertainties are of a technico-economic nature. This applies whether it is a matter of monocrystals, organic molecules having pharmacological properties or very high-grade glass. It is not yet possible to predict, on the basis of an analysis of the various factors involved in the competition between earthbound and in-orbit technologies, whether this type of activity will be viable. This is very clear from the widely differing assessments supplied by experts, which vary between a few hundred million dollars per year -nothing in relation to the current levels of space expenditure (several tens of billions of dollars) -and upwards of a hundred billion dollars per year, which would turn this activity into the major driving force behind the development of space technology. We find ourselves in an inherently unstable situation, which has implications for national strategies and makes it impossible to know with a reasonable degree of certainty what the future will bring.

With regard to antiballistic weapons, i.e. the in-orbit development and deployment that may result from President Reagan's strategic defence initiative, all that can be said is that the same sort of technological uncertainty as that hanging over microgravity is compounded by political uncertainty, of which a fairly accurate idea can be gained from reading the press.

As we consider the possible future course of space technology -i.e. the feasibility of progressing through intermediate stages to attain a higer level of activity and eventually perhaps one at which production activities could be transferred to Space on a very large scale -we find ourselves in a position of uncertainty that bears more than a passing resemblance to the position at the beginning of the sixties. People thinking about the future of space then were far from sure what it might hold and wondered whether it was in fact reasonable to invest so much effort in preparing for it.

However, there is a practical question to be asked: what are the technologies that will need to be mastered in order to move into a new phase of space technology, making it possible to contemplate production activities? I have my own personal view on this, one obviously influenced by what I have read on the subject, by the ideas of others therefore. In my opinion, the most important technology to master involves occupying the surface of an atmosphere-less celestial body and working under the conditions prevailing there. It is no use imagining that items for space can continue to be produced, as they are today, from materials torn from the deep gravitational well formed by the Earth's mass; sources closer to Space will have to be found.

Whether to choose the Moon or, for example, Phobos, a satellite of Mars, is a tactical matter. I think the choice should be determined above all by the desire to take an irreversible step, to achieve permanent mastery of a technology and avoid becoming involved again in an Apollo-type operation, i.e. a "demonstration". The time for demonstrations in Space is past; the choice of a celestial body on which to install a permanent experimental base should be guided, as far as possible, by reason. From such a base, it will be possible gradually to master the techniques involved in travelling between the celestial body and Earth and in working on the former's surface. I personally regard the Moon as the better choice, though that remains very much open to discussion.

Supposing that the technological system were to expand into the solar system rather than confining itself to the earth's biosphere, would this result in the colonisation of Space? A question that cannot fail to grip the imagination; various distinguished thinkers -I mention Gerard O'Neil and Freeman Dyson just by way of example- have produced some very worthwhile work on the subject.

At the risk of disappointing, I consider this question to be premature. The industrialisation of Space is an essential prerequisite to its colonisation. It is a stage that must be gone through, and colonisation is inconceivable unless industrialisation has already taken place. Thus, the prospect of the colonisation of Space does not, for the time being, have any direct bearing on what I believe should be the main aim of study of the future of Space, namely the formulation of as coherent and rational an overall strategy as possible.

Reference

[1] Preliminary Design for an Experimento-World-Circling Spaceship

LIFE SCIENCES IN SPACE

Heinz Oser
Life Sciences, Microgravity Office
Directorate of Earth Observation and Microgravity
European Space Agency
8-10, rue Mario Nikis, 75738 Paris Cedex 15, France

Life Sciences Research in space starts on Earth. Life on our planet has developed over many millions of years under favourable conditions, notably the atmospheric gases composition, pressure, temperature, radiation and under the constant pull of earth's gravity. Amongst these physical factors, also the force of gravity has shaped life on Earth.

Early scientific description and documentation of the influence of gravity on life forms has been put forward by Darwin, mentioning the geotropic response of plants, i.e. plants put in a horizontal position were bending over a period of time back into a vertical position perpendicular to earth's force. No valid concept of the meaning of this phenomenon was suggested at that time, let alone the fact of descriptive underlying mechanisms or processes.

Only as late as in the 1930's, life scientists made some valid predictions about potential changes of organisms in weightlessness with respect to morphology and function, at a time when spaceflight allowing for the compensation of gravity forces due to 'free fall conditions' was still unthinkable for the public world.

Only with the onset of manned space flight approximately 30 years ago, life science research in space began. Initially dedicated to research and experiments about survival in space, this time saw a number of different species exposed to spaceflight, foremost of all dogs, primates, and ultimately man himself. And it was not only weightlessness, there were the high acceleration forces during launch and reentry and the exposure to the unshielded cosmic radiation which were of concern to those undertaking spaceflight. Later, almost any form of life was exposed to space conditions ranging from bacteria, to plants, amphibia, insects, rodents on an increasing scale of complexity: yet most of this research was dedicated to look and see what happens. And it is no surprise that for almost two decades into spaceflight, this was not considered as a 'proper science'.

Only now, over the last few years, there was a considerable change in the conceptual world of life scientists: still small in their number, they nevertheless realised that gravity force can be a variable in their research and thus a <u>tool</u>.

Since most of the data obtained were related to man's well being,

J. Schneider and M. Léger-Orine (eds.), Frontiers and Space Conquest, 19–24.
© 1988 by Kluwer Academic Publishers.

and to the development of countermeasures for proper adaptation, a considerable 'data base' has been built up over the years. But despite some well controlled experiments, this must still be seen in many cases as 'field studies'.

However, today we can afford to dedicate this science to more fundamental research, and thus probing into the question: what does gravity mean, and what role does it play in processes fundamental for life on Earth?

Let us have look on living cells, small but nevertheless utmost complex units of life. ESA has developed a small laboratory, for use on Spacelab, called BIORACK, allowing for well controlled and reproduceable experiment procedures, including 1g control centrifuges during orbital flight; and it is only just a little more than a year ago that first evidence was produced for an influence of gravity on cellular level:

- Proliferation, the process for cell development, growth and multiplication seems to be increased under weightlessness, and suggestions are made that this might be due to less energy expenditure than on Earth. The question arises as to whether indeed gravity acts as an inhibitor for an explosive multiplication of life on Earth.

- Differentiation, the process by which cells assume a distinct function, e.g. release of hormones, antibodies, develop into malignant cells, or keep external bodies under control, seems to be markedly decreased under weightlessness, suggesting perhaps a gravity influence for the adaptation and multitude of life forms on our planet Earth.

- Conjugation, the process in which the exchange of genetic material occurs, between cells, is distincly increased under weightlessness, suggesting not only that these fine bridges are not broken by the pull of gravity, but also that gravity may influence even subcellular processes. Moreover, and widely speculated, would this indicate again an earthbound protective mechanism for almost 'uncontrolled' of life forms on Earth?

- Sedimentation of subcellular structures with a higher density than their surroundings (e.g. amyloblasts in plant cells) is not only activated by gravity but may also explain the gravitropic response of plants, a feature which is vital for Earth's vegetation and our survival on this planet; (Have you thanked your plant today? asks a bumper sticker seen in the U.S., and anyone on our planet should ask himself this vital question daily);

- Biorhythm, at least on unicellular organisms, persist in weightlessness with a remarkably stable periodicity and amplitude, suggesting the origin of endogenic rather than exogenic terrestrial Zeitgebers, a sort of a built-in-clock. Almost all life forms display a rhythm, and periodicity may vary from seconds to days, or even years, but the deep meaning of it is by far not understood. Is there any universal principal behind?

- Embryogenesis, and oogenesis, the process in which life develops and takes shape, and which was followed up with insect eggs, indicated even a potential synergism between cosmic radiation and weightlessness, the meaning of it in no way understood and best of all just an observation as of today, however remarkable.

Let us have a look now on man in space, the other extreme of complexity in life forms. In general, there is a rapid and good adaptation to the weightlessness environment, and a more cumbersome re-adaptation to earth's gravity. Although on a short term basis (days) there are significant alterations in e.g. massive fluid shifts from the limbs to the upper torso and into the head (puffy faces), which is primarily due to the abolishment of hydrostatic forces acting on Earth, there is soon an equilibrium reached which persists for the remaining spaceflight. And although there are initially some significant alterations in the neuro-sensori system, probably due to conflicting information by man's inbuilt accelerometer, the otoliths in our inner ear, and the other gravity dependent stimuli, again within a couple of days an equilibrium is reached suggesting a re-interpretation of external cues and supporting man's extraordinary rapid adaptation to an alien environment never over exposed to for millions of years. Moreover, once adapted, man thoroughly enjoys his three dimensional freedom in hitherto unknown feeling. And almost exclusively, visual information seems to dominate the world of space travellers, being able to completly ignore the up and down dimensions so used to on Earth.

However, looking into longer space missions, say for months or years, there is valid concern that a more thorough adaptation may take place than wanted, or even man is made for: muscles, in particular our anti-gravity muscles, become hypothrophic, and decalcification processes start in our skeleton, suggesting that not only there is no need for it in weightlessness but also running into the danger of potentially irreversible processes. It should be emphasized here that neither the underlying mechanisms have been understood until now, nor that valid countermeasures could be developed, both of which are still key obstacles for truly long term manned space missions. And quote the Soviets, who have extensive experience, complaining about the difficulty to avoid any adaptation of the human body to weightlessness, a seemingly eternal fight against the difficult return to Earth conditions. It almost seems that we interpret today true adaptive changes as adverse reactions yet they are, in principle, benign in their nature and rather support man's great dream of leaving even mother Earth. And having talked about robotics versus man in space: truly this is not an either or question, but when man is needed, his brain is still by far superior (for quite a time to come) when the combination of his intelligence, communication and motoric skills are required: just this little region in the central brain, called gyrus praecentralis, predominantly coordinates all movements of speech, fingers, arms and eyes in a manner and dexterity unmatched by any robot, however big or sophisticated.

And in the spirit of this philosophical colloquium let me draw a more provocative conclusion on the above: what would man be like in space, borne over many, many generation? For sure there will be no need for gravity supporting structures -and the adaptation points in that direction- like the weight bearing part of the skeleton and anti-gravity muscles both of which alone account for approx. 70% of our body mass; and there is no real need for legs other than toes would function in primates just to stabilize one's position in a three dimensional environment; apart from the need for a large brain and eyes with a corresponding change in the hierarchy of processing 'out-side' information, there is no need for gravity sensing organs and systems, as much as man already does not possess the feel for speed but rather the change of speed, i.e. acceleration. Moreover, cosmic radiation resistant tissues, and adaptation to low metabolism for long periods of time, as well as an inbuilt propulsion system, would be highly desired assets. If one would put a matrix on the above requirements together and skims through all the species on our Earth, the octopus would rank very high on this scale of (almost) perfect space travelers.

How about the future of life sciences in space?

As much as Darwin pointed to an evolution through natural selection of random heriditary fluctuations, as much there will be later on a trend to search for a more universal principle. And already now, yet not widely known, there is a field of research on a however modest basis, called exobiology, which means research to reach a more universal understanding of the phenomenon life which, until now, has only meant life on Earth. It includes the understanding of the origin, evolution and distribution of life in the universe, by means of investigating cosmic evolution of biogenic compounds, to look into the early evolution of life as such and to search for extraterrestrial life forms and intelligence.

In our solar system, the comets, the chrondritic meteorites and the interplanetary dust give evidence of a dynamic chemistry which prevailed at the beginning, during condensation of the solar nebula. It is likely that in the dynamic system of the solar nebula, several condensation processes operated simultaneously. Examples are thermo-chemical reactions, photochemical reactions driven by UV-radiation of the young sun, radiation chemical reactions initiated by cosmic rays, reactions under influence of lightning discharges or shock waves.

The chain of cause and effect to the appearance of life on Earth, to its spread and multiformity still puzzles us. Has gravity favoured its development, or was there rather an adaptation to it? Theories and speculations are based on information mainly from: geological and astronomical observations, meteorite analyses, sophisticated experiments under conditions simulating those of the primitive Earth, fossile findings and comparative studies of present day organisms. Environmental fluctuations may have served as stimuli from the self-organisation of polymers to functional complexes until membrane formation provided for phase separation from the external environment = last step of individuation which is considered as a first living cell on Earth, and which in layman terms may be considered as proper definition of life.

If one condensed the development of life on Earth -from the most primitive living organisms ever defined as life as far as man himself -to a one year scale, starting from 1 January to 31 December, one would see a startling evidence of the dimension time. From first of January until about September virtually nothing happened what could be remotely described as life- and some biochemists even include here certain forms of clay. Only towards the end of September appeared the formation of pre-biotic material. Towards October/November one would gradually find appearance of most primitive life forms e.g. bacteria, and it would not be until December to see, for the first time, higher organisms like primitive algae or other unicellular structures. By mid-December, in our terms rapid evolution, even higher organised multicellular organisms developed and it was not until end of December that primates appeared just to see at the last day in December, virtually before midnight Homo Sapiens on Earth.

Yet all of the above has not remotely contributed to answers about the question: where does life, as far as we understand it, really come from? And this will remain one of the deepest questions for mankind, probably not being capable of understanding these universal principles, but rather seeking solutions and believe in theological and philosophical aspects. However, looking into the scientific aspects, already in 1903 Arrhenius put forward the theory of natural panspermia which means "seeds of life" are transported through space by radiation pressure until they reach a planet which supports development and growth. This model of panspermia does not, however, explain the mode and location of the origin of life somewhere in the universe.

If one looked through a microscope on a living cell, seeing the nucleus, floating or sedimenting denser matter in a fluid environment, seeing the fragile tiny but vital cell membrane, and taking inside created or from the outside supplied energy into account, one would immediately realize the startling similarity with our Earth if looked at by a telescope from far away: its fragile thin atmosphere, so vital for Earth, resembling a membrane, and in further analysis the solid and liquid components, including the nucleus, and the careful balance between inside created and outside supported energy; in other words, a truly living cell in space; it is just a question of dimension.

Very first simple model experiments in this direction have been realised on Spacelab missions, and are planned for the first EURECA mission.

But it is just now within the foreseeable future, as far as mankind can grasp it and project ahead for generations to come, that the emphasis on life sciences research in space will thus greatly shift to this extraordinary field of science, namely to investigate and look for such a UNIVERSAL PRINCIPLE.

ECHO PHILOSOPHIQUE A L'INTERVENTION DE H. OSER

Maria Villela-Petit
chargée de recherche en philosophie
C.N.R.S. Paris
59 rue Lhomond 75005 Paris, France

L'analyse biomédicale des données expérimentales moissonnées au cours des séjours humains dans l'espace extraterrestre (que vient de nous proposer H. OSER), me conforte dans la conviction qu'une des tâches actuelles de la philosophie consiste à penser l'homme à partir d'une réhabilitation de la question du corps. Or cette question, me semble-t-il, est inséparable de celle de la Terre.

En assignant à la philosophie la tâche de penser à nouveaux frais la question du corps, je n'inaugure aucune préoccupation inédite, mais je m'inscris dans le sillage de la phénoménologie, dont un des mérites à été d'orienter la réflexion philosophique vers le dépassement d'un dualisme de type cartésien. Mais il importe maintenant de prolonger la phénoménologie, en la critiquant si nécessaire, en vue d'articuler plus finement la question du corps et celle de la Terre. Husserl lui-même nous en ouvrait la voie lorsqu'il faisait place dans son oeuvre à ce que j'appellerai un géocentrisme au second degré -non plus celui d'une cosmologie préscientifique, mais celui qui a pour lui une vérité d'ordre phénoménologique. Et cela, curieusement, dans un texte prémonitoire où la possibilité d'une vue de la Terre à partir d'un autre astre était déjà envisagée [1]. Le point en question ici, et pour l'indiquer en quelques mots, est le suivant : nous pouvons nous représenter la Terre comme un corps plus ou moins sphérique tournant autour du soleil. Une telle représentation, que nous disons copernicienne, est la seule qui soit scientifiquement, objectivement vraie. Ajoutons qu'aujourd'hui avec les voyages dans l'espace une telle représentation a reçu un remplissement intuitif. Nous avons pu voir la Terre d'en haut, les astronautes et nous autres, restés sur Terre, grâce aux transmissions télévisées et aux photographies. Grande banalité, dirait-on. Et pourtant, ce que Husserl a voulu mettre au jour c'est que sous cette représentation de la Terre comme un corps se cache une couche plus originaire de l'expérience que nous en faisons. A ce niveau plus originaire de notre expérience, la Terre se donne comme sol, comme Terre-sol. En tant que telle, en deçà du mouvement ou du repos, elle est ce sur quoi et à partir de quoi nous faisons l'expérience des corps qui sont sur terre, en repos ou en mouvement, ainsi que des corps célestes. Il y a donc une priorité phénoménologique de notre expérience de la Terre comme _sol_ sur notre représentation de la Terre comme _corps_. Il en va pour la Terre, comme pour notre propre corps. Je peux bien objectiver mon corps, l'imagi-

J. Schneider and M. Léger-Orine (eds.), Frontiers and Space Conquest, 25–29.
© 1988 by Kluwer Academic Publishers.

ner en dehors de moi, et pourtant aucune distance objectivement mesurable ne me sépare de lui ; c'est lui qui définit mon ici, cet ici à partir duquel tous les autres corps m'apparaissent. Le parallélisme n'est cependant pas absolu, car alors que je ne peux quitter mon corps qu'en imagination, je pourrais éventuellement quitter la Terre, la voir d'en haut, du dehors. Mais que ce soit sur un avion, sur une navette ou sur un autre astre, la question du sol se posera toujours. Or, d'après Husserl, ce fragment de sol, serait unifié transcendantalement avec la Terre-sol [2].

Nous savons, quant à nous, que la question du sol n'est pas aussi simple que le croyait Husserl, lequel n'avait pas pu envisager la question de l'apesanteur, disons de la microgravité.

Quoi qu'il en soit, le sens de la Terre pour nous est à chercher en deçà même de la question du sol. La Terre est d'abord et avant tout -à un niveau d'originarité qui échappe au regard phénoménologique, puisqu'à ce niveau tout regard déjà la présuppose-, matrice et nourrice de toute vie. De cette vie qu'à juste titre nous appelons terrestre. En exergue à ce qui demande à être pensé au titre de cette appartenance de notre vie à la Terre, nous pourrions évoquer ici une des proclamations du Zarathoustra de Nietzsche "*Plus loyalement, plus purement discourt le corps en bonne santé, parfait et bien carré, et son discours concerne le sens de la Terre*" [3].

Laissons de côté ce qui relève dans cette proclamation d'une idéalisation du corps. Un tel corps parfait n'est pas celui des hommes en chair et en os que nous sommes. Appelé par Nietzsche comme le corps du surhomme, il est davantage conçu sur le modèle de la statuaire grecque. Nietzsche devait, d'ailleurs, y songer expressément, ainsi qu'en témoigne l'adjectif "carré". Cette épithète, qui disait pour les Anciens Grecs la perfection, était justement celle dont on se servait pour qualifier les oeuvres du grand sculpteur classique Polyclète. Il reste néanmoins que la question est posée : le corps discourt, parle du sens de la Terre. Autour d'un tel noyau de pensée, une méditation pourrait se développer prenant tout aussi bien en considération les corps souffrants, les corps affamés, du fait, par exemple, de l'inclémence du soleil, lorsque, cessant de faire germer la Terre, l'astre-roi la rend aride. Sans oublier ce qui dans les corps chétifs renvoie à l'incurie ou à l'injustice des hommes dans leur partage de la Terre, ou/et, en dernière analyse, à la fragilité même de la vie.

L'idée-force ici pour nous est que le corps porte dans sa structure même les traces de son appartenance à la Terre.

Toutefois, le projet autour duquel nous sommes ici réunis n'est-il pas justement celui d'envisager pour l'homme un habitat autre que la Terre, habitat provisoire en un premier temps, mais peut-être demain définitif (?), au moins, pour une petite parcelle de l'humanité ? Ne sommes-nous pas, voyage après voyage, en train d'étudier, de mettre à l'épreuve les possibilités du corps humain de façon à lui donner les moyens de survivre et de vivre dans des conditions extraterrestres ? Qui le nierait ?

Mais ne nous laissons-nous pas leurrer par les mots. Les conditions extraterrestres que connaissent nos astronautes ne sont extraterrestres que dans un sens bien relatif. Certes, la question de l'apesanteur ou de la microgravité altère considérablement la question du sol, et par conséquent tout ce qui s'y rattache au niveau même du vécu corporel, mais qu'en serait-il de nos astronautes s'ils ne disposaient pas de ces "nourritures terrestres", que sont l'oxygène, l'eau, les aliments riches en teneur protéique et vitaminique ? Que l'on puisse obtenir de l'eau sur la Lune en y transportant des bonbonnes d'hydrogène, ou obtenir de l'oxygène sur Mars en y cassant des molécules de dioxyde de carbone, nous devons l'envisager, mais cela ne transformera pas la planète ou notre satellite en jardins, et, en attendant que ces déserts fleurissent, des hommes ne pourront y survivre, (ou, comme c'est, en un premier temps, plus probable, en station orbitale) que grâce aux provisions terrestres et à une sorte de reproduction mimétique du milieu terrestre. Et puis, il y a à compter avec le facteur chronologique. Combien de temps nos astronautes peuvent-ils rester hors du milieu terrestre sans préjudice grave pour leur santé, sans un appauvrissement de leur sensibilité ? Appauvrissement qui serait inéluctable si des hommes devaient être engendrés en vaisseau spatial ou dans une station construite sur une planète désertique. Et l'appauvrissement du langage qui en résulterait ?

Or malgré le côté sommaire de ces remarques, elles peuvent servir d'instance critique chaque fois que nous lisons, sous la plume d'un scientifique, l'annonce que, dans un avenir prévisible, on va pouvoir couper le cordon ombilical qui nous relie à la Terre-Mère. L'annonce enthousiaste de cette prochaine libération d'une Mère abusive, puisqu'empêchant ses enfants de devenir adultes, outre qu'elle cache des ressentiments peut-être inavouables, trahit tout simplement le fait que l'on s'est laissé abuser par la métaphore usée de la Terre-mère. Mais les mots aussi sont dangereux, eux aussi peuvent nous enchaîner. Ainsi tous ces discours qui ne savent saluer l'astronautique qu'au prix d'un déclassement, bien naïf, du terrestre, se rattachent, à leur insu, à de bien vieilles traditions. Ils font écho en sourdine à des formules de genre : "C'est vraiment une grande misère de vivre sur la Terre." Dans un cas comme dans l'autre on prétend répondre ainsi à l'appel du ciel. Mais alors que le spirituel du XVème siècle entendait cet appel comme un appel transcendant : celui du supraterrestre, le scientifique à l'aube du XXIème siècle l'entend comme un appel immanent : celui de l'extraterrestre. Le supraterrestre, auquel l'on n'avait accès qu'après la mort, ayant été remplacé par un extraterrestre, qui semble parfois comporter un déni phantasmatique de la mort.

Et pourtant l'homme est terrestre. Cette affirmation va de plus en plus s'enrichir d'un contenu positif au fur et à mesure qu'avanceront les études faites sur les astronautes, soumis à ces véritables situations expérimentales que constituent les séjours dans des navettes ou dans des stations orbitales. Nous allons de mieux en mieux comprendre au moins au niveau biomédical comment notre appartenance à la Terre est inscrite dans notre corps, comment sa forme en dépend.

Cela dit, je m'empresse d'ajouter, pour ne pas laisser mon propos prendre une tournure unilatérale, que l'homme est également le seul vivant terrestre à écouter l'appel du ciel, à s'ouvrir à une véritable contemplation, encore que les animaux soient, eux aussi, sensibles aux variations du jour et de la nuit et à la lueur des astres. Une telle capacité de s'ouvrir au ciel ne serait-elle pas constitutive pour l'homme de son humanité ? Et n'est-ce pas c'est dans l'angle de cette ouverture que la science et la philosophie sont nées ? Il reste qu'aucun appel du ciel n'échappe à une profonde méprise s'il vient à impliquer un refoulement du terrestre. Un tel refoulement qui, aujourd'hui, prend la forme d'un refus de ménager la Terre et de respecter les vivants qu'elle porte, s'avère déjà comme suicidaire.

Pour conclure, sans vouloir conclure, c'est-à-dire en laissant les questions ouvertes, je vous invite à méditer ceci : et si l'homme ne pouvait trouver sa véritable place dans le cosmos que s'il apprenait à marier en lui l'appel du ciel et la fidélité de la Terre ?

Notes

[1] La communication de P. Hadot a montré l'ancienneté et la portée éthique de cette image de la Terre vue d'en-haut. Toutefois, après la révolution copernicienne, la représentation de la Terre comme pouvant être vue du ciel acquiert un autre sens, tout en perdant son sens spirituel et éthique, que suggérait déjà la "hauteur" de la vue. Il s'agit désormais de voir la Terre comme un astre parmi d'autres, sans statut spécifique propre. V. là-dessus Descartes qui, dans Les Principes de la Philosophie, 3ème Partie, § 8, écrit : *Que la Terre étant vue du ciel ne paraîtrait que comme une planète moindre que Jupiter et Saturne*". C'est par rapport à la révolution copernicienne déjà pleinement acquise qu'il faut comprendre le "renversement" qu'opère à son tour E. Husserl, lorsqu'il essaie de penser le primat phénoménologique de la Terre.

[2] E. Husserl, Recherches fondamentales sur l'origine phénoménologique de la spatialité de la nature (L'arche originaire Terre ne se meut pas), trad. fr. par D. Franck, in Philosophie, n° 1, Paris, Janvier 1984, p. 3-21.

[3] Cf. Nietzsche, Ainsi parlait Zarathoustra, texte établi par G. Colli et M. Montinari, trad. fr. M. de Gandillac, Gallimard, 1971, p. 44.

Intervention

Jean Heidmann

Je voudrais insister sur le fait que l'unicité de la vie intelligente dans notre galaxie n'est qu'une pure possibilité. Tout d'abord je rappelle qu'un tel énoncé ne fait que limiter le problème puisque l'on sait pertinemment qu'il existe au moins cent milliards de galaxies. Après trente ans de débats spéculatifs on ne sait absolument rien sur le nombre de civilisations intelligentes dans notre galaxie. La question est maintenant abordée scientifiquement et observationnellement par cette nouvelle branche de la science nommée "bioastronomie", c'est-à-dire la recherche de la vie dans l'univers menée par le côté "astronomes". Plus particulièrement, par de nouvelles technologies résultant d'une recherche-développement menée par la NASA, des récepteurs radio à dix millions de canaux d'écoute simultanés (alors que les récepteurs courants de la radioastronomie ne disposent que d'un millier de canaux simultanés) vont être mis en oeuvre pour un programme d'écoute d'une demi-douzaine d'années, et permettre d'envisager ainsi observationnellement la mise en évidence de l'existence ou non de la vie intelligente dans l'univers.

LA TERRE VUE D'EN HAUT ET LE VOYAGE COSMIQUE
LE POINT DE VUE DU POETE, DU PHILOSOPHE ET DE L'HISTORIEN

Pierre Hadot
Professeur au
Collège de France
11, place Marcelin Berthelot, Paris, France

Dans une lettre à Schiller, datée du 12 mai 1798, Goethe écrivait ceci : "*Votre lettre m'est parvenue hier lors d'une lecture de l'Iliade, poème auquel je reviens toujours volontiers, parce qu'on y est élevé, comme dans une montgolfière au-dessus de tout le terrestre et qu'on se trouve véritablement dans cet espace intermédiaire dans lequel planent en tous sens les dieux*" [1]. Dans ce texte, Goethe établit une liaison entre un événement, qui était pour lui encore tout récent et qui l'avait beaucoup impressionné, et une représentation archaïque, celle du vol des dieux. Pour la première fois en effet, en 1783, l'homme s'était arraché à la pesanteur terrestre et s'était élevé dans les airs. C'est à cette expérience que Goethe se réfère pour faire comprendre l'impression qu'il éprouve en lisant Homère. Il est vrai que la poésie d'Homère nous élève en quelque sorte de terre en nous faisant voir d'en haut les choses humaines. En effet l'auditeur archaïque ou le lecteur moderne des poèmes d'Homère, qui est en imagination le spectateur des événements décrits par le poète, apprend par celui-ci qu'il y a des êtres qui, du haut des airs, ou du haut des montagnes, assistent aux événements qu'il raconte. Ce sont les dieux, spectateurs privilégiés, qui, d'ailleurs, descendent parfois des cieux pour se mêler des affaires humaines. Et ainsi, tout naturellement, l'auditeur ou le lecteur fait coïncider son regard avec celui des dieux. Il s'élève, comme dit Homère au chant V de l'Iliade "*dans l'étendue qui sépare la terre du ciel étoilé avec les chevaux d'Héra, qui*, nous dit-il, *dévorent autant d'espace en un instant que peut en embrasser le regard d'un guetteur épiant, d'un point élevé, la surface de la mer*" (V, 771). Ainsi le vol des dieux est aussi rapide que le vol d'un regard. Et au début du chant XIII de l'Iliade, avec Zeus, Homère et son lecteur regardent toute l'étendue terrestre, le pays des Thraces et celui des Mysiens et de bien d'autres nations, ou bien, avec Poséidon, "*s'émerveillant*, nous dit Homère, *de la bataille et de la guerre*", ils se perchent sur le plus haut pic de Samothrace et prennent pitié des Grecs, au moment où ils sont défaits par les Troyens.

On peut généraliser. Ce regard porté d'en haut vers la terre, c'est le regard du poète lui-même, le regard de la poésie, le regard que la poésie nous donne sur les choses : "*La vraie poésie*, dit Goethe, *se*

J. Schneider and M. Léger-Orine (eds.), Frontiers and Space Conquest, 31–39.
© 1988 by Kluwer Academic Publishers.

reconnaît au fait que, comme un évangile profane, elle sait nous libérer des pesanteurs terrestres qui nous accablent en nous donnant à la fois la sérénité intérieure et le plaisir extérieur. Comme un ballon, continue Goethe, *elle nous élève dans les régions supérieures avec le lest qui tient à nous et elle fait apparaître sous nos yeux, révélés et dénoués, les labyrinthes terrestres qui nous paraissaient inextricables*" [2].

La vraie poésie, selon Goethe, nous emporte donc, elle aussi, au-dessus de la terre comme le ballon de Montgolfier, mais aussi comme ces ailes que Dédale avait fabriquées, pour s'échapper du labyrinthe qu'il avait lui-même inventé et dans lequel Minos l'avait enfermé.

Pourquoi la poésie, en nous arrachant mentalement à la pesanteur terrestre, a-t-elle le pouvoir de nous apporter la sérénité intérieure, et de nous arracher aux labyrinthes où nous sommes enfermés ? Nous le comprendrons mieux tout à l'heure. Mais, pour l'instant, il nous faut surtout retenir que Goethe établit une liaison entre l'élévation de l'esprit au-dessus de la terre et la sérénité intérieure, autrement dit que pour lui l'élévation produit un effet moral, mieux encore, une sorte de transformation spirituelle. Cela nous permet de mettre à part ces textes de Goethe, parmi des centaines d'autres textes, qui, depuis l'Antiquité jusqu'à nos jours, ont pour thème la description d'une élévation au-dessus de la terre et d'un voyage cosmique et qui peuvent revêtir, d'ailleurs, les formes les plus variées, comporter les mises en scènes les plus diverses, se présenter comme des voyages imaginaires, au-dessus de la terre ou dans l'espace, dans l'univers ou au-delà de l'univers, avec le corps ou sans le corps, avant la mort ou après la mort, en rêve ou pendant la veille. Lorsque Goethe compare la poésie à une élévation au-dessus de la terre, il a une représentation très particulière de ce vol mental. Il conçoit en effet la poésie comme un exercice spirituel pratiqué par le poète et par son lecteur. Il ne s'agit donc absolument pas d'un rêve de vol qui s'imposerait au poète ou d'une rêverie à laquelle il se laisserait aller, mais d'une démarche voulue et consciente destinée à procurer à l'homme la sérénité intérieure.

Goethe est, en cela, l'héritier de la philosophie antique, dans laquelle cet exercice joue un rôle capital dans toutes les écoles. Déjà Platon décrivant le philosophe dans le Théétète, 173, nous dit que c'est par son corps seul qu'il habite dans la Cité, mais que "*sa pensée, pour qui toutes les affaires humaines ne sont que mesquineries et néant, dont elle ne tient pas compte, promène partout son vol, sondant ce qui est sous terre, mesurant ce qui est sur terre, étudiant la marche des astres sur la voûte qui domine le ciel, explorant en totalité toute la nature de chacune de ces réalités, sans jamais redescendre à ce qui est immédiatement proche*". L'esprit du philosophe, dit ailleurs Platon (Républ., 486a) contemple la totalité du temps et de l'espace, c'est pourquoi il n'a en lui aucune bassesse, aucune mesquinerie, mais au contraire il possède l'élévation de la pensée, c'est pourquoi aussi il regarde la vie humaine comme une chose sans grande importance, et il n'a pas peur de la mort.

Dans presque toutes les écoles philosophiques, on retrouve la pratique de cet exercice, destiné à provoquer la sérénité de l'âme.

Cet exercice spirituel a un double effet : d'une part, en plongeant l'esprit dans l'immensité du cosmos, il lui procure un sentiment de bonheur, de plénitude, de sérénité et, d'autre part, en faisant regarder d'en haut les choses de la terre, il permet de les juger à leur juste valeur, il délivre l'esprit des passions, des soucis, des craintes et des ambitions. Ces deux aspects de l'exercice se retrouvent dans toutes les écoles. Pour les Epicuriens, c'est surtout l'infinité de l'espace et des mondes qui provoque un sentiment d'enthousiasme : "*Puisque l'espace*, dit Lucrèce, *s'étend à l'infini au-delà des murailles de ce monde, l'esprit cherche à savoir ce qui se trouve dans cette immensité où il peut plonger ses regards, aussi loin qu'il veut et où il peut s'envoler d'un essor libre et spontané*" (II, 1044-1047). "*Les murailles du monde s'envolent. Je vois dans le vide immense naître les choses... La terre ne m'empêche pas de distinguer tout ce qui, sous mes pieds, s'accomplit dans les profondeurs du vide. A ce spectable, je me sens saisi d'un frisson de plaisir divin*" (III, 16-17 et 27-29). Comme dit l'épicurien Métrodore : "*Souviens-toi que, né mortel, avec une vie limitée, tu t'es pourtant élevé, par les recherches physiques, jusqu'à l'éternité et l'infinité des choses, en voyant ainsi l'avenir et le passé*". Le sage épicurien jette en quelque sorte sur l'infini de la réalité le même regard que les dieux bienheureux tels que les conçoit Epicure, étrangers au monde, contemplant avec sérénité l'infinité des choses.

Dans la tradition platonicienne et stoïcienne, on retrouve le même sentiment fondamental, mais formulé dans la perspective d'une autre conception de la physique. Les philosophes, selon Philon d'Alexandrie (De special. leg., II, § 44), "*aspirant à une vie de paix et de sérénité, contemplent la nature et tout ce qui se trouve en elle, ils explorent attentivement la terre, la mer, l'air, le ciel,... ils accompagnent par la pensée la lune, le soleil, les évolutions des autres astres, errants ou fixes, leurs corps restent sur terre sans doute, mais ils donnent des ailes à leurs âmes pour que, s'élevant dans l'éther, elles observent les puissances qui s'y trouvent, comme il convient à des hommes qui sont devenus citoyens du cosmos*". "*L'âme*, écrit le stoïcien Sénèque, (Quaest.Natur., I,7-12) *atteint la plénitude et l'achèvement du bonheur que peut atteindre la condition humaine, lorsqu'elle gagne les hauteurs et parvient jusqu'à l'intérieur du sein de la nature. elle se plaît à planer au milieu des astres... Arrivée là-haut, elle s'y alimente et grandit : libérée de ses entraves, elle revient à son origine*".

L'astronome Ptolémée, chez qui l'on rencontre des vestiges des doctrines platoniciennes, stoïciennes et aristotéliciennes, exprime lui aussi, dans une épigramme qui lui est attribuée avec quelque vraisemblance l'impression qu'il éprouve d'être associé à la vie divine, lorsqu'il se plonge par la pensée dans les espaces célestes : "*Je le sais, je suis mortel et ne dure qu'un jour. Mais quand j'accompagne, dans leur course circulaire, les rangs pressés des astres, mes pieds ne touchent plus terre, je vais auprès de Zeus lui-même me rassasier d'ambroisie, comme les dieux*" [3]. Ces vers ont été imités, on le sait, par Tycho Brahe en 1574 et par Kepler [4] en 1596.

Quant à l'empereur stoïcien Marc Aurèle, contemporain de Ptolémée, il écrit dans ses Pensées (VII, 47) : "*Embrasser du regard les courses des astres, comme s'ils nous emportaient dans leurs révolutions et avoir constamment dans l'esprit les transformations des éléments les uns dans les autres. De telles représentations purifient des souillures de la vie terrestre*".

Cette Pensée nous permet d'entrevoir la raison pour laquelle Goethe affirmait que la poésie nous élevait au-dessus de la terre. On peut dire en effet que les philosophes de l'Antiquité pratiquent la physique comme un exercice spirituel [5]. Je veux dire par là que, pour Platon, pour Epicure, pour les stoïciens, la contemplation du monde et de l'espace cosmique a, avant tout, une finalité morale. Epicure par exemple écrit (Ratae Sententiae, § 11) : "*Si nous n'étions pas troublés par nos craintes concernant les phénomènes célestes et la mort... nous n'aurions pas besoin de faire de la physique*" ou encore (Epître à Hérodote, 635 : "*Je recommande d'appliquer une constante activité à l'étude de la physique, considérant que c'est cette activité qui procure le plus de sérénité dans la vie*". Le Timée de Platon est lui aussi un exercice spirituel [6], puisqu'il nous invite (90) à redresser les mouvements de notre esprit en les mettant en harmonie avec les mouvements des astres dans le cosmos.

Chez Platon, chez Epicure, chez les stoïciens, on peut dire que l'exercice spirituel de la physique consiste à se délivrer du point de vue conventionnel, "humain trop humain", pour voir sa propre vie et les choses humaines dans l'immensité du cosmos, dans la perspective de la nature universelle, dans l'infinité de l'espace et du temps, et plus spécialement pour considérer toutes choses dans la perspective des grandes lois de la nature, comme Marc Aurèle, dans le texte que nous venons de citer, évoquait l'universelle métamorphose. Ces grandes lois de la nature ce sont par exemple pour Marc Aurèle, outre cette transformation de toutes choses les unes dans les autres, l'unité vivante du cosmos, l'harmonie et la correspondance de toutes choses entre elles. Et la physique, considérée comme exercice spirituel, se donne pour tâche de replacer la vie humaine dans la perspective de ces lois universelles. C'est la raison pour laquelle, je pense, Goethe conçoit la poésie comme un exercice qui consiste à s'élever en esprit au-dessus de la terre. Goethe a en effet une conception très particulière de la poésie. Pour lui, la poésie précisément est une sorte de physique au sens où nous venons de la définir, c'est-à-dire un exercice spirituel qui consiste à regarder les choses d'en haut, dans la perspective de la nature universelle, dans la perspective de ces grandes lois de la nature que sont pour Goethe, non seulement l'universelle métamorphose et l'unité de toutes choses, mais les deux processus universels de la polarisation et de l'intensification : Polarität und Steigerung que Goethe se plaît à observer aussi bien dans la nature que dans la vie humaine. Un tel exercice procure la sérénité intérieure.

J'ai dit tout à l'heure que l'exercice spirituel qui consiste à s'élever en pensée au-dessus de la terre et à s'envoler dans le cosmos avait un double effet : d'une part en plongeant le philosophe dans l'immensité de la nature, il lui donnait un sentiment de bonheur et de sérénité, d'autre part, en lui faisant regarder d'en haut vers la terre, il lui permettait de juger les choses humaines à leur juste valeur.

Nous venons d'étudier le premier thème. Quant au second, nous l'avons déjà vu s'esquisser dans la République et dans le Théétète de Platon. Nous le retrouvons, richement orchestré, dans les écoles philosophiques les plus diverses. Pythagore, dans les Métamorphoses d'Ovide (XV, 147), s'écrie : "*Je veux m'élever parmi les astres, je veux quitter le séjour de la terre immobile,... de là-haut je verrai les hommes errant à l'aventure, privés de raison, et tremblant de peur à l'idée de la mort*". C'est d'en haut également, que le philosophe, selon Lucrèce (II, 9) porte ses regards sur les autres hommes, et qu'il les voit errer de toutes parts, cherchant au hasard le chemin de la vie. Et, selon le stoïcien Sénèque (Questions Naturelles, I, 8) c'est lorsqu'elle a fait le tour du cosmos, lorsqu'elle est au milieu des astres, que l'âme du philosophe jette du haut du ciel un regard dédaigneux sur la terre minuscule et qu'elle se dit : "*C'est là ce point que tant de nations se partagent par le fer et le feu ! Combien sont risibles les frontières que les hommes mettent entre eux !*" Du haut du ciel, les armées de la terre lui paraissent des fourmis. Et c'est lorsqu'elle est là-haut ajoute Sénèque, que l'âme du philosophe se rit du luxe des riches, de leurs mosaïques, de leur or, de leurs portiques.

Chez Marc Aurèle, ce thème prend une forme très réaliste. C'est un véritable exercice de l'imagination dans lequel l'empereur s'efforce de voir, de se représenter la vie de l'humanité dans le présent, l'avenir et le passé comme le montrent les Pensées VII, 48 et IX, 30 que je vous cite : "*A qui veut parler des hommes, il faut observer les choses terrestres comme s'il se trouvait en quelque lieu où, d'en haut, il regarderait vers le bas : rassemblements de foules, armées, travaux des champs, noces, divorces, naissances, morts, brouhaha des tribunaux, déserts, diversité des moeurs des peuples barbares, fêtes, lamentations, marchés, ce pêle-même et finalement l'ordre harmonieux des contraires*". "*Regarder d'en haut : rassemblements de foules par milliers, ces fêtes innombrables, toutes ces navigations dans la tempête ou le beau temps, et toutes ces diversités des êtres qui naissent, qui vivent ensemble, qui meurent*".

Cet effort pour regarder d'en haut permet donc de contempler le panorama total de la réalité humaine sous tous ses aspects sociaux, géographiques, sentimentaux, et de les replacer dans l'immensité cosmique et dans le fourmillement anonyme de l'espèce humaine sur la terre. Vues dans la perspective de la nature universelle, les choses qui ne dépendent pas de nous, les choses que les stoïciens appellent indifférentes, la santé, la gloire, la richesse, la mort, sont ramenées à leurs vraies proportions.

Le thème du regard d'en haut, quand il prend cette forme particulière : l'observation des hommes sur la terre, paraît appartenir plus particulièrement à la tradition cynique. Nous le retrouvons en effet, très richement mis en scène, par un contemporain de Marc Aurèle, le satiriste Lucien, qui précisément a été fortement influencé par le cynisme. Dans le dialogue de Lucien intitulé : L'Icaroménippe ou l'homme qui s'élève au-dessus des nuages, le cynique Ménippe raconte à un ami comment, découragé par les contradictions des philosophes au sujet des principes derniers de l'univers, il a décidé d'aller voir lui-même dans le ciel ce qu'il en était. Il s'est donc ajusté des ailes pour voler, l'aile droite d'un aigle et l'aile gauche d'un vautour. Il s'élance ainsi dans la direction de la Lune. Lorsqu'il y est parvenu, il voit d'en haut la terre toute entière et, comme le Zeus d'Homère, nous dit-il, il observe tantôt le pays des Thraces, tantôt le pays des Mysiens (nous retrouvons le passage de l'Iliade dont nous avions parlé tout à l'heure) et même, s'il le veut, la Grèce, la Perse et l'Inde : ce qui le remplit dit-il d'un plaisir varié. Et il observe aussi les hommes : "*Toute la vie des hommes m'est apparue*, déclare Ménippe, *non seulement les nations et les cités, mais tous les individus, les uns naviguant, les autres faisant la guerre, les autres en procès*". Et il n'observe pas seulement ce qui se voit en plein air, mais aussi ce qui se passe dans les maisons où chacun se croyait bien caché. Remarquons en passant que c'est là le thème du célèbre roman du XVIIIe siècle "Le diable boiteux" écrit par Lesage. Après une longue énumération d'exemples de crimes, d'adultères, qu'il voit se commettre à l'intérieur des maisons, Ménippe résume ses impressions en parlant de pêle-mêle, de cacophonie et de spectacle ridicule. Mais le plus ridicule de tout à ses yeux est de voir les hommes se quereller pour les limites d'un pays, car la terre lui apparaît minuscule. Les riches s'enorgueillissent de bien peu de chose. Leurs terres, dit-il, ne sont pas plus grandes qu'un des atomes d'Epicure, et les rassemblements des hommes ressemblent au grouillement des fourmis. Ayant quitté la lune, Ménippe voyage à travers les étoiles pour atteindre Zeus et il s'amuse à constater le ridicule et les contradictions des prières que les humains adressent à Zeus. Dans un autre dialogue, intitulé Charon ou les surveillants, c'est le passeur des morts, Charon, qui demande une journée de congé pour aller voir à la surface de la terre ce que peut être cette vie sur terre que les hommes regrettent tant lorsqu'ils arrivent aux Enfers. Avec Hermès, il entasse donc plusieurs montagnes les unes sur les autres pour pouvoir mieux observer les hommes. Nous retrouvons alors le même genre de description que dans l'Icaroménippe et chez Marc Aurèle : navigations, armées en guerre, procès, travailleurs des champs, activités multiples, mais vie toujours pleine de tourments. "*Si, dès le début*, dit Charon, *les hommes réalisaient qu'ils sont mortels, qu'après un bref séjour dans la vie, ils doivent en sortir, comme d'un rêve et laisser tout sur cette terre, ils vivraient plus sagement et mourraient avec moins de regrets.*" Mais les hommes sont inconscients. Ils sont comme les bulles produites par un torrent qui s'évanouissent à peine formées.

Ce regard d'en haut sur la vie terrestre des hommes revêt, nous l'avons dit, une forme propre au cynisme. Ce qui le prouve, entre autre, c'est que le dialogue intitulé Charon a pour titre en grec Episkopountes, "Ceux qui surveillent". Or le philosophe cynique considère que son rôle est de surveiller les actions des hommes, qu'il est une sorte d'espion qui guette les fautes des hommes et les dénonce. C'est ce que dit Lucien lui-même. Le cynique est chargé de surveiller les autres hommes, il est leur censeur, il observe leur comportement comme du haut d'un observatoire. Et les mots episkopos, kataskopos, "surveillant", "espion", sont attestés dans la tradition antique pour désigner les cyniques [7]. Ce regard d'en haut, pour eux, est destiné à dénoncer le caractère insensé de la manière de vivre des hommes. Il n'est pas indifférent que, dans l'un des dialogues de Lucien, ce soit Charon, le passeur des morts qui regarde ainsi d'en haut les choses humaines. Il les voit en effet dans la perspective de la mort. Regarder d'en haut, c'est aussi regarder les choses humaines dans la perspective de la mort : c'est cette perspective qui donne le détachement, l'élévation, le recul indispensables pour voir les choses telles qu'elles sont. Le cynique dénonce la folie des hommes qui, oubliant la mort, s'attachent passionnément à des choses, le luxe et le pouvoir, qu'ils seront obligés d'abandonner inexorablement. C'est pourquoi le cynique appelle à rejeter les désirs superflus, les conventions sociales, la civilisation artificielle qui sont pour les hommes une source de troubles, de soucis, de souffrances et il les invite à revenir à une vie simple et purement naturelle.

Pour Lucien, nous l'apprenons dans son petit livre : "Comment il faut écrire l'histoire", le regard qui se porte d'en haut vers les choses humaines n'est pas seulement celui du philosophe mais aussi celui de l'historien. Ou plus exactement le regard de l'historien doit être celui d'un philosophe : c'est-à-dire courageux, impartial, étranger à tout pays, bienveillant pour tous, ne donnant rien ni à la haine ni à l'amitié (§ 41). Et cette attitude doit se traduire dans sa manière de raconter les faits. Il doit être, nous dit-il, comme le Zeus d'Homère qui tantôt jette les yeux sur le pays des Thraces, tantôt sur celui des Mysiens. Nous retrouvons donc pour la troisième fois ce regard divin homérique qui se pose d'en haut sur la terre, mais cette fois pour y trouver le modèle de l'impartialité qui doit s'exprimer dans la structure même du récit, grâce au point de vue élevé où l'historien se place (§ 49). Ici la vision d'en haut apparaît comme la condition de l'objectivité de l'historien, de son impartialité. C'est ce que les modernes appelleront le point de vue de Sirius. Renan écrira par exemple en 1880 : "*Quand on se place au point de vue du système solaire, nos révolutions ont à peine l'amplitude de mouvements d'atomes. Du point de vue de Sirius, c'est encore moins*" [8]. Se placer au point de vue de Sirius, c'est, encore une fois, pratiquer un exercice spirituel de détachement, de distanciation, pour atteindre à l'impartialité, à l'objectivité et à l'esprit critique.

Disons-le en passant, je crois, que sur ce point, il ne faut pas chercher des excuses dans les théories de Raymond Aron sur les limites de l'objectivité historique. Celles-ci nous invitent surtout à prendre conscience des dangers et des illusions qui nous guettent. Mais elles ne dispensent personne de l'effort moral pour se libérer de la partialité et de la passion.

Dans toute cette tradition philosophique, l'effort pour s'éloigner de la terre grâce à l'imagination était un moyen pour se libérer intérieurement d'une manière de voir trop individuelle ou trop anthropomorphique et pour voir toutes choses dans la perspective du cosmos, en rapport et en proportion avec le cosmos. En un certain sens, l'image du vol était un symbole de la force de l'esprit humain qui est un pouvoir de dépassement, un pouvoir pour aller toujours au-delà. On peut donc dire que la tradition de l'Occident s'est longuement préparée en imagination au voyage cosmique effectif et a essayé d'entrevoir à l'avance les transformations qu'il pourrait entraîner dans la représentation du monde, dans la conscience des individus, dans la représentation que l'humanité se fait d'elle-même. Nous avons vu notamment comment cet exercice spirituel pouvait amener le philosophe à dénoncer la vanité et l'injustice des inégalités sociales et l'absurdité de la guerre, comment, grâce à lui, l'homme se concevait lui-même comme un citoyen du cosmos, comment il éprouvait, en le pratiquant, le sentiment d'une transfiguration, d'un dépassement de la condition humaine, qui le délivrait de la crainte de la mort et lui procurait la paix et la sérénité intérieures.

J'ai donc parlé du passé et c'était là l'essentiel de mon propos. En conclusion, je me hasarderai à parler de l'avenir pour dire d'ailleurs que, du point de vue psychologique et moral, il est totalement imprévisible. Ceux qui, au XIXe siècle, prophétisaient l'avenir de l'humanité au XXe siècle avaient-ils prévu qu'à la fin du XXe siècle, précisément au début de la conquête de l'espace, on verrait renaître avec une puissance inouïe les intégrismes religieux ?

Maintenant que les voyages cosmiques ont commencé, les hommes ont-ils changé, sont-ils devenus plus pacifiques, plus sereins, plus humains ? Ne peut-on pas dire plutôt que l'homme emmène dans l'espace la terre elle-même, non pas la Terre partie du cosmos, mais la terre symbole de l'humain trop humain, et des mesquineries humaines et que l'on peut craindre que le cosmos ne devienne bientôt le théâtre des absurdes guerres de religions qui continuent en notre XXe siècle à déchirer l'humanité. La conquête de l'espace risque de donner seulement un champ plus vaste à la folie humaine.

Les voyages cosmiques, je le crois, ne dispenseront donc jamais du voyage cosmique intérieur, celui que les philosophes, les poètes, les sages ont osé entreprendre, chacun à leur manière, celui dont parlait le regretté G. Friedmann, dans son beau livre La Puissance et la Sagesse lorsqu'il écrivait : "*Prendre son vol chaque jour ! Un moment qui peut être bref, pourvu qu'il soit intense. Chaque jour un exercice spirituel... Sortir de la durée. S'efforcer de dépouiller tes propres passions... S'éterniser en se dépassant*" [9].

Références

[1] Briefe, Hamburger Ausgabe (= HA), t. 2, p. 344
[2] Goethe, Dichtung und Wahrheit, livre XIII, HA, t. 9, p. 580

[3] *Anthologie Palatine*, IX, 577 et voir A.J. Festugière, *La Révélation d'Hermès Trismégiste*, t. I, Paris, 1944, p. 317, auquel j'emprunte la traduction

[4] Jean Kepler, *Le Secret du monde*, trad. Alain Segonds, Paris, 1984, p. 1 et p. 233, n. 2 (bibliographie)

[5] P. Hadot, *Exercices spirituels et philosophie antique*, 2e éd., Paris (Les Etudes augustiniennes), 1987

[6] P. Hadot, "Physique et Poésie dans le Timée de Platon", *Revue de Théologie et de Philosophie*, t. 115, 1983, p. 113-133.

[7] Par exemple Epictète, *Entretiens*, III, 22, 24.

[8] E. Renan, Oeuvres complètes, t. II, Paris, 1948, p. 1037

[9] G. Friedmann, *La Puissance et la Sagesse*, Paris, 1970, p. 359.

THE FUTURE AMERICAN, SOVIET AND EUROPEAN SPACE PROGRAMMES

R.M. Bonnet
Director of ESA's Scientific Programme
8-10 rue Mario-Nikis, 75738 Paris Cedex 15, France

This paper gives a brief description of the long and very long term programmes of the world's principal space agencies. Details of those agencies' short and medium term programmes are available in the relevant documents and reviews. Our main concern here, however, is with the prospects for space exploration in the 21st century. The documents on which our study is based are listed in Table 1 and are all of American or European origin. We have no material concerning the Soviet programme. While it is true that the Russians are prepared to describe their short and medium term plans at bilateral and multilateral meetings, it is nevertheless difficult to obtain reliable information going beyond the immediate future. In attempting to form a long-term picture, we must each rely on our own analysis, using extrapolations and assumptions based on past, present and imminent Soviet projects.

Table 1

- Pionnering the Space Frontier US National Commission on Space	Mai 1986
- Major Directions for Space Research National Academy of Sciences	1987
- The Crisis in Space and Earth Science, a Time for a New Commitment NASA Advisory Council	Novembre 1986
- ESA Long Term Plan ESA/C(84)46 Space Science Horizon 2000 ESA SP 1070	

Of all the documents listed in this Table, "Pioneering the Space Frontier" is the first of its kind to lay down a guideline for the American space programme for the next fifty years. Whatever view is taken of its content, it alone offers us a vision of the future. It is an original and enthralling piece of work, produced by a groupe of men and women concerned to provide their country with a long-term policy that will prevent concern with immediate goals leading down the wrong path or to a dead end. They deserve our gratitude for having helped to map out, for the rest of the world, the main stages in the human occupation of space in the 21st century. The strategy outlined

J. Schneider and M. Léger-Orine (eds.), Frontiers and Space Conquest, 41–50.
© 1988 by Kluwer Academic Publishers.

is a possible one. Moreover, it is not incompatible with those adopted by Europe and the Soviet Union. Indeed, the latter are currently implementing various stages of the American plan! This state of affairs, which the facts compel us to recognise, does not appear fortuitous, since in the area there seems to be only one possible line of action, whose variants are of merely secondary singificance. We shall return to this point in the section dealing with the Soviet and European programmes.

Which way is human civilisation heading?

The first question to be asked is what defines the 20th century. The answers to that question -for there must necessarily be more than one- will depend on the kind of person you ask.

Obviously, the 20th century is characterised by the advent and application of the major theories of physics, and by the entry of biology and medicine into the domain of science. It is also the century of the technology explosion - that of the motor car, aeroplane, atom, space rocket and computer. And it is the century of experiment with extreme social systems.

I would personally be inclined to define the 20th century as the century of electronic communication, by which term I denote all the techniques and processes that transform a physical, biological or social phenomenon, or an event of any kind, into an electrical signal that can be transmitted to any point on the planet irrespective of its distance from the place of occurence of the phenomenon in question. By such means almost every human being can be put within sight or hearing of any point on Earth, and thus enabled to follow any event, whether political or scientific, in real time. Today information from whatever source is virtually instantaneously exchangeable, once detected and transformed into electrical signals, by telephone, television, artificial Earth satellite or data distribution network. It is this that will make the 20th century the century of truth, at least in respect of effects if not of causes.

We can now pose the second question, although proposing an answer to it is a riskier matter. What will define the 21st century? The safest answer is probably a further question.

Will the 21st century not be the one in which the planet Earth, entwined in its skein of communication networks, proves too cramped for the human race? Matters are already clearly moving in that direction: in politics, where global relations are very decidedly determined by the shift to large geopolitical units; in regard to ecological issues and human destruction of the natural environment (pollution of the seas, Chernobyl, saturation of the geostationary orbit); in science too, since the Earth no longer provides a long enough base for the large interferometers that will revolutionise the astronomy of tomorrow; and so on.

It is not too difficult to predict that this trend will accelerate in the 21st century and will probably be the factor leading the human race to break free of the Earth and to give freer rein to its aspirations, as well as to an expansion more justified than ever before by the limitations which the natural dimensions of our planet impose on us. Indeed, the need to expand will doubtless be a greater factor than curiosity about the unknown, since our observation techniques will have enabled us to analyse in detail the surface of our sister planets, the asteroids and the moons of Jupiter and Saturn, and to discover that planets like the Earth and Mars are legion throughout the universe.

In the 21st century humanity will seek to measure itself against its new dimension, the dimension of space. This prospect, I believe, is what gives the work of the National Commission on Space, chaired by former NASA administrator Tom Paine (which offers the human race a way out of the predictable dead-end for which it is heading), and it is to this prospect that we now turn our attention.

The long and short term prospects held out by the American programme

The two hundred pages or so of the Paine Commission report provide us with a full description of a possible programme of space exploration and settlement to be pursued by future American governments. In this plan, little is left to chance; each step, each component, is consistent with all the others and with the project as a whole.

THE AIMS

The first aim is the colonisation of space, i.e. of the near solar system, or te put it another way, avoiding the unpleasant connotations of the term "colonisation", the establishment of the first human settlements in space. The next aim is the establishment of a system of free and independant communities among the new worlds, communities that would be set up on the Moon or on Mars, thus providing potential new resources for humanity through extraction of the riches locked beneath the surface of those planets. The third aim is to achieve and maintain American leadership in space for the next fifty years. It is not a question of "escaping" from the Earth - the point is to ensure a continuous American presence on the Moon and Mars. A project of this kind is no easy thing to carry through, nor does the Paine report make any attempt to gloss over the difficulties.

THE DIFFICULTIES

First come the physical difficulties. The fact that gravity at the Earth's surface is six times greater than that of the Moon and three times that of Mars means that we are at the bottom of a tremendous gravity well from which any attempt to escape to the highways of interplanetary space will be enormoulsy complicated and costly. This becomes a major obstacle as soon as large masses are to be transported. Hence the very logical idea of putting such masses together bit by bit in Earth orbit, where a barrier of 8 km/s has already been crossed and only another 3 km/s is needed to break free of the Earth's gravitational pull. An hence also the importance of a permanently occupied space station where the assembly operations can be carried out. Similar considerations point to the great advantages of low-gravity bodies, which are much easier to land on and easier to leave. For that reason the asteroids, above all Phobos and Deimos, the two asteroids held captive by Mars, are highly attractive "spaceports", both in terms of the exploitation of their own resources and as staging posts.

Then there are the financial difficulties. At the present cost of space transport, a space colonisation project on the scale envisaged would require an unreasonable level of funding that is inconceivable, whatever is at stake, unless systems are redefined in such a way as to bring down costs drastically. That is why transport of astronauts and transport of cargo logically require two different systems, one of which, used infrequently and providing absolute reliability, will necessarily be more expensive than the other, which will need to be used much more often but not have to be as reliable. This means developing new specialised vehicles for each function.

There are very definitely technical difficulties, which require maximum use to be made of existing concepts that have been fully mastered - the space shuttle and space station. Nevertheless, long-duration voyages over great distances will inevitably require the development of new electrical and nuclear propulsion systems. In support of such a programme, the Paine report calls for an approximately threefold increase in NASA's advanced technology research budget.

There are also the biological and physiological difficulties involved in keeping human beings alive in space during long interplanetary voyages and after they have reached their destination. Here there are three difficulties of the first order. First, the need to counter the known effects on the human organism of several months' stay in space. Second, the need to study the possible effects of the conditions of life one Mars or on the Moon, where gravity is only one third and one sixth respectively of that on Earth; for this purpose it is proposed to investigate the long-term effects of such conditions in space, using a variable artificial gravity facility. Third, the need to protect human beings against the lethal effects of cosmic radiation, which means fitting the spacecraft with bulky and heavy screening and, once the astronauts have landed on the Moon or Mars, building shelters out of the materials available there. Finally, feeding the astronauts requires the use of recycling systems and veritable space greenhouses, a prototype of which is already operational in Arizona.

In conclusion, mention must needs be made of the political difficulties. The plan proposed can be adopted only as the result of political resolve on the part of the American government. But if that is the case, how can compatibility be ensured with the corresponding Soviet, European or indeed Japanese and Chinese plans? Given that international cooperation appears, from both the political and the financial point of view, to be the key that could turn the epic enterprise of the 21st century from a dimly perceived vision into a reality, how can this approach be reconciled with the idea that the proposed plan can be acceptable to the American government only if it ensures American leadership in space?

INFRASTRUCTURE

The infrastructure proposed consists of a number of elements intended to link the Earth with the Moon and then with Mars. The first elements is a "Highway to Space" leading from Earth to the Space Station. Hardware would be carried along it by cargo vehicle. A new type of spacecraft, half rocket half plane, capable of reaching Earth orbit but taking off and landing like an aeroplane, could be an economic way of transporting astronauts. Construction of such a vehicle is recommended as an immediate priority.

Next are the "spaceports": a space station in Earth orbit, the two moons of Mars, Phobos and Deimos, which also contain natural resources of value to the future Martian colonies, and a lunar port that could be located at the Lagrange point between the Earth and the Moon, where the gravitational pulls of the two bodies exactly cancel each other out, with the advantage that little energy would be required to approach or leave it. The spaceports would be linked by a "Bridge between Worlds" consisting of transfer vehicles docking at and taking off from the spaceports.

The final elements in the plan, which it is envisaged will take fifty years to complete, are a Moon base and a Mars base. The bases would form the nuclei of the future human colonies. While their initial purposes would be scientific, they would subsequently serve to provide resources for establishing the future colonies, which would have to find their means fo survival on the spot. The Moon's soil is by mass 40% oxygen, shile some areas of its surface are rich in hydrogen of solar origin. The creation of mini-biospheres is thus conceivable. In the case of Mars, matters are even simpler, since water, in the form of frozen particles, is present in the subsoil.

To carry this plan through and set up the infrastructure in question would require a carefully thought-out strategy. The Commission envisages three stages:

1. Exploration of sites and natural spaceports, involving study of the moons of Mars and observation and systemic study of the various constituent elements of the solar system: planets, moons, asteroids and comets. Space survival requires the solution of numerous biological and physical problems in microgravity or reduced gravity conditions. Scientific research and space science are therefor given a key role in this strategy. Space science in particular, with its enormous power of fascination, will make it possible to draw in new blood - the generation of scientists,

engineers and technicians who will in the end be responsible for carrying this ambitious task to its conclusion.

2. Prospection, followed by establishment of the planetary bases that will lay the foundations of the future space colonies. It is during this phase that the sites most propositious for human settlement will be identified, more particularly in terms of mineral riches, oxygen and hydrogen reserves in the case of the Moon, and water in the case of Mars. Finally, it is during this phase that human beings will build shelters to protect themselves against the hostile environment, and then their future living units.

   The timing of the Space exploration, as seen by the National Space Commission can be summarized by three main steps:

   - human return to the Moon in 2005, preceded by robotic return in 2000

   - Moon base in operation in 2020

   - a robot on Mars in 2012, followed by human beings in 2015 and a permanent base in 2028.

3. The last phase is when humans cease to be pioneers and explorers and settle for good on one or other of these new worlds. But this is not envisaged as happening less than fifty years from now.

Colonisation of the Moon precedes that of Mars for purely logical reasons. It is much easier to establish a settlement on our natural satellite because of its proximity (maximum flight duration 3 days) than on the Red Planet, which takes 6 to 8 months to reach. The Moon is thus a virtually obligatory intermediate stage.

Implementation of this strategy depends on the availability and development of a number of systems. For that reason the Paine Commission attaches prime importance to the development of the space station, to robotics, and to the development of new launch vehicles - both a spaceplane (Orient Express) for the transport of astronauts, and automatic cargo vehicles for transporting equipment and materials. Another necessary step along the road is the study of the long-term effects of stays in reduced gravity on the human organism, which will provide knowledge of the conditions for long-duration flights and the survival of humans on the Moon and Mars. For this purpose studies and technological development of all the relevant systems - automatic flight, propulsion, life-support, food supply, work, etc. - are essential and need to begin without delay if the time schedule for the various stages of the plan is to be met.

Finally, a programme of such scope calls for a major funding effort that would reflect the political commitment of governments to a course of action that is manifestly decisive for the future of humanity. The funding involved, however, would not be out of all proportion, since it would represent, in dollar equivalents, no more than half of the maximum annual budget ever allocated to the Apollo programme, and an almost constant fraction of American GNP. Nothing extravagant, then, provided the political will is there. And that means not only the will to embark on the plan, but also the will to give continous support to all stages of it, whatever problems arise en route. This is perhaps where the greatest difficulty lies, since the American system is not yet capable of ensuring the permanence of such an undertaking, if only because the federal budget is subject to annual decisions and can be called into question every year.

Soviet prospects

As already mentioned, we unfortunately do not have the equivalent of the Paine report in the case of the Soviet space programme. It is moreover very hard to see how, and at what level, a similar long-term planning exercise could be undertaken in the context of management of the Soviet programme. No doubt the Soviet Academy of Science has a key role to play in a possible coordination of all the facets of the programme, but the Academy is not alone, because the programme involves a dominant military element that is apparently beyond its jurisdiction. The logical conclusion is that coordination is a matter for the highest political authorities, and this reflects a fundamental aspect of the Soviet programme, namely the total support of the political authorities, which makes it an unchallengeable component of Soviet policy.

Without a document, a Soviet "Paine report," all we can do is to observe and analyse the main features of Russian space policy and make some extrapolations that may be bold but are no more than logical. Now, what we in fact observe is that nothing in the Soviet programme runs against the recommendations of the Paine Commission and that, quite to the contrary, each element of the programme, or item of infrastructure involved, complies closely with those very recommendations.

The Mir space station, for example, which has been operational since 1986, is clearly the nucleus of a larger station on which Soviet cosmonauts and those of other countries will stay for long periods, thus providing a permanent Soviet presence in Earth orbit. Mir, like its Salyut predecessors, is equipped with life-support systems providing conditions for virtually autonomous human survival; in addition, miniature biospheres that are fully recycling and autonomous have, although they are still at the experimental stage, been in operation on Soviet spacecraft for several years.

As far as planetary exploration is concerned, absolute priority is given to Mars and its moons. In fact, in the Soviet Union the name Phobos is more often associated with a mission in preparation than with the Martian moon itself. The mission involves sending two spacecraft towards Mars in July 1988 to make a detailed study of the surface and chemical composition of Phobos, an asteroid caught in the gravitational field of the Red Planet. A further step is to be taken in 1992 or 1994 with the "Mars 92" mission, which will land a probe on Mars itself and carry out a soil analysis while an aerostatic station studies the planet's winds and atmosphere. The mission will also put an artificial satellite in orbit around Mars, one of the main tasks of which will be to observe and identify potential landing and prospection sites for a subsequent more ambitions mission to be undertaken before the turn of the century.

The purpose of the subsequent mission would be to land a double robot on Mars that, like the Lunakhod in the 70s, would explore several sites, take soil samples and transport them to a spacecraft that would carry them back to Earth. The spacecraft in question, which we can suppose to be some kind of rocket, would rendez-vous with an orbiting satellite that would play a role very similar to that of the "Bridge between Worlds" described in the Paine report. The samples would travel in this manner from Mars to Earth, where they would be taken on board the Mir station in order to avoid contamination of the Earth by any micro-organisme hidden in the Martian subsoil.

Of plans beyond the year 2000, nothing is yet known, but it is not unreasonable to suppose that the combination of the two visible components of present Soviet strategy, to wit long-duration flights and exploration of Mars by means of automatic probes, is the prelud to landing cosmonauts on Mars in the course of the 21st century. Indeed, the Russians already have launchers that could do this. Their Proton rocket, which has now been in use for about twenty years, is a powerful and reliable launcher whose series production suggests that it has the advantage of economy, despite (or perhaps because of) the fact that the technology involved, which was ahead of the field twenty years ago, is now old hat. In addition, the Russians are known to be building a space shuttle and a new heavylift launcher, and are also talking about a return to Moon exploration.

One cannot help but notice that everything is happening as if the Russians had already begun to implement the recommendations of the Paine Commission! Might it be that there is only one viable strategy for humankind's use and settlement of space? At all events, that hypothesis would not be refuted by European plans.

European prospects

Apart from programmes for scientific exploration of the universe, the Europeans have no general space project going beyond the end of the century. The European Space Agency and its Member States are nevertheless engaged in developing a space infrastructure fully in line with that envisaged by the United States and the Soviet Union and designed to achieve European autonomy in the human occupation of space.

The Europeans are envisaging participating alonside the Americans in the construction of the Space Station. European participation, known as the Columbus programme, consists in the provision of two pressurised modules, one attached to the station and the other free-flying but capable of accomodating astronauts, plus two platforms for scientific instruments only, one in polar orbit and the other co-orbiting with the station nucleus.

Europe is also planning to develop a cargo vehicle, Ariane V, capable of lifting 15 tonnes into Earth orbit in one go. This vehicle will also be able to launch Hermes, the European spaceplane, which will carry four astronauts to the Space Station in the years 1995 to 2000. Beyond this phase, the development of a new transport system along the lines of the American "Orient Express" is being studied. The study is based on two possibilities: the British Hotol concept of a horizontal take-off rocket-powered spaceplane, and the German Sanger project of a shuttle-type spacecraft lifted initially on the back of an aeroplane.

In addition to this infrastructure, ESA intends to develop scientific, meteorological and telecommunications programmes fully capable of competing with those of the United States and Soviet Union. The inescapable fact remains, however, that Europe does not have, and will not have for a long time to come, that capacity for the heroic that sustains the programmes of the two superpowers and of which Neil Armstrong, Sputnik I and Yuri Gagarin are symbols. For Europe is not a superpower, and its space programme suffers from the intrinsic difficulty of imbuing an entity so culturally, intellectually and economically rich with the political spirit that it still so sorely lacks and that would enable it to play a full part in the epic conquest of space, rather than having constantly to cling to the superpowers' coat-tails.

Conclusions

It would thus seem that only one strategy exists for the use and settlement of space, and that it is common to the American, Soviet and European programmes. It is not simply an intellectual exercise: the strategy, developed to a greater or lesser extent by its different protagonists, sets out objectives that far transcend mere opportunism. What is at stake is the future of human civilisation, whatever the political regime with it may locally be identified. While we may be encouraged by the existence of a common approach, it is a matter for concerne that it rests in each case on primarily nationalistic considerations.

The Paine report, which tacitly assumes that the future space exploration and settlement programme will be carried out in an international context, implicitly limits the non-American input to 25%, the remaining 75% presumably being intended to ensure American leadership. How can this be reconciled with both the Soviet strategy, which is also on international cooperation, and the European strategy, which aims at greater autonomy vis-à-vis the United States? The politicians will clearly have to come up with a responsible reply to this question in the not-too-distant future.

One is also struck by the contrast between the immensity of the philosophical issues with which we are concerned here and the rather limited, not too say parochial, ambitions of the great powers. After all, we are not talking about colonising the universe, but simply about achieving, over the next fifty years, a permanent human presence on the Moon and Mars, which are no more than suburbs of the Earth.

At the present time, the technical difficulties are a major problem for the development of a more ambitious space settlement programme, but there is no shortage of all kinds of other problems. No doubt it will take a long time to get round them all, and many decades may pass before they are solved. Nevertheless, it must be acknowledged today that it is on space that our future, and probably our very survival, depends, even if our ambitions are modest. (Some, on the contrary, may consider them unrealistic).

In these circumstances the international programmes described in this paper provide us with a means of immediate entente - and perhaps even detente. It is now or never! Let us seize the opportunity before it is too late.

THE "HIGH FRONTIER" CONCEPT IN PERSPECTIVE

Dr. Roberto Pinotti
Sociologist
Futuro srl, Florence, Italy

G.K. O'Neill's pioneer studies and hypotheses on XXIst century space colonies still seem the stuff of science fiction dreams to many. His space colony has the shape of a cylinder with round-off ends. Large mirrors, angling out from one end, reflect sunlight into the interior through window panels which constitute half the surface area of the cylinder; people live in the cylinder interior. Cables extending to outer space can connect this colony with another one. The cylinders would be rotating to provide gravity and would act as gyroscopes. With two of these structures spinning in opposite direction and linked together, the resulting system would behave as if there were no spinning at all and could easily be made to track the sun. The first colony, according to this concept, should be 600 feet in diameter, rotating at 3 revolutions per minute to give normal gravity. The colony would then be a cylinder a mile long. But since there could be trouble if the spin were faster than one rpm, this meant the colony could not be a cylinder but had to be redesigned into the shape of a torus over a mile in diameter with people living on the inside of it, 400 feet wide: a sort of replica of the classic space station in Stanley Kubrik's 2001: A SPACE ODISSEY. In any case, as O'Neill commented on, when he testified in the U.S. Senate hearings on January 19, 1976, "*it is natural for most people, and particularly for reporters and art directors, to become preoccupied with two features of orbital manufacturing... One is, 'Where is it (the space colony) going to?' and the other is, 'what is it going to look like?'. I think the proper answer to the first question is 'in an orbit high enough so that it almost never gets eclipsed (in Earth's shadow, i.e. shaded from the sun), and to the second, it will be a rotating pressure vessel, containing an atmosphere, with sunshine brought inside with mirrors'. Beyond that, any further detail is almost certain to be wrong. For that reason, among others, I think it's unnwise to get personally identified with particular designs. I'm for whatever works best, and it's too soon yet to be sure what that will be*".

As we move into outer space with space stations, lunar bases and space colonies, the crew requirements will become more and more varied, and so will the problems of dealing with people. In any case, one thing is certain: such new conditions urge future social models for man in space and beyond Earth.

As far as we know, this is surely a real challenge for social

J. Schneider and M. Léger-Orine (eds.), Frontiers and Space Conquest, 51–58.
© 1988 by Kluwer Academic Publishers.

sciences, and it won't be easy to answer to all the questions generated by man's activity in more or less confined situations and in lethal environments. Nevertheless, it is evident that psychologists and sociologists must begin to face all this as far as possible. There are many who think that living in outer space will only cause stress, depression, boredom and thoughts of home in future astronauts. Others, on the contrary, hope that such different life conditions will, in some magic way, produce a better human being, since surely space colonization will be a good opportunity to sweep out the old and pour in the new.

Since the second half of the XIXst century science fiction presented more or less convincing literary hypotheses of man's future life. And philosophy has been always facing the problem of suggesting new and better life schemes for man in ideal cities, from the Greek "polis" to today's "metropolis". After the Noelithic Revolution human communities tried to develop their new permanent habitat: the city. Apart from its dimensions, the city is something more than an artificial product of man. In fact, it is the place where our species act and interact, and the basis for nearly all our culture and history. It can't properly be likened to anything else in nature, because it is something far beyond an animal species' integrated group, as in the case of bees, ants, termites, and other social insects. It is our race's test tube without which art, science and progress could not exist, and mankind's key to its future.

This social product of man had different faces during our history: dreams and hopes confronted the reality of facts and life, in a sort of everlasting antithesis; nevertheless all this had a deep influence in our evolution. Plato's "Republic" and his description of Atlantis in "Timaeus" and "Critias", Thomas More's "Utopia", Francis Bacon's "New Atlantis", Tommaso Campanella's "La Città del Sole" are just some of the works suggesting a better life scheme for man. Of course we had utopian but also dystopian outlooks, like in modern science fiction, but in any case it is clear that we are going to project all our problems and contradictions in outer space as well. What about the final result up there?

G.K. O'Neill was rather careful about this and did not make definite judgements about the possible social structures of his future space colonies. But certain values are evident in his proposals; in fact, there is no mention of conflict, hate, depression or stress on personal, bureaucratic, economic, political or ideological levels. As he wrote, his colonies are designed for a "*comfortable life... enlarged human options... breaking through repression*".

Of course this optimistic point of view was not enough, and so the team who worked on the NASA Space Colony Study in 1975 as a consequence of O'Neill's pioneer studies gave some consideration to possible, different kinds of social organization.

At first the NASA team envisioned one community that had majority rule as the basis of democracy, with competition as the key of its development or "progress". This community was described as "hierarchical and homogeneous"; possible differences were regarded as accidental or merely inconvenient, and diversity would be considered abnormal and undesirable; in other words, "common earth-type deviance".

Such a society might feel that differences create conflicts and therefore would prefer to organize everything and everyone in the light of maximum efficiency. Of course it was not all. Another type of society foreseen in the 1975 NASA study was one in which independence is raised to a high level of virtue both for the individual and the whole group as such. Such a community would consider self-sufficiency as the highest form of human behavior. As Robert M. Power wrote in his The Coattails of God, "*a community like this might be loosely based on the precepts of individuality and creativity discussed in the novels and essays fo Ayn Rand*". Last but not least, a third type of social order considered in the NASA study was one in which heterogeneity is considered a source of flexibility, evolution and survival as well. As a consequence of this, rather than thinking in terms of the greatest efficiency, such a society might think more in terms of choosing and matching people in order to obtain reasonable results. A community so structured might regard competition as quite useless but co-operation very helpful. It is evident that in this case the overall idea is more Eastern than Western. Diversity is possible only in a general harmony. This is just the spirit of the Russian space program. Inside their MIR space station orbiting Earth the Soviet cosmonauts act not as individuals but as parts of an unique, gigantic effort. And from their particular viewpoint they are perfectly right, after all.

Space was and is surely worth risking their lives for all the astronauts and cosmonauts. Space crews looking at the Earth found that they were being drawn into a new frame of mind. Of course it would be difficult for man to spend several months floating weightless in a new environment where local verticals shifted from room to room and where he was first one way up and then another, and then come out with the same perspectives as when he went in; at the very least, he would be receptive to new points of view. The astronauts knew in advance, of course, that the world doesn't have boundaray lines between countries marked on it like on a library globe, but they were nonetheless rather surprised when they saw directly from space that there were no kinds of physical divisions on the Earth's surface as well as between peoples. "*In no way could we on Earth, or any group of people, or any Country, consider ourselves isolated: we are all in this together...*" SKYLAB astronaut Gibson said after his mission. "*When you're up there, you see the Earth as one unit, you see the sun as a star, and you can see all the other stars out there, and you realize that the universe is quite big*".

National, cultural or religious differences disappear in space. "*At first I tried to see on the Earth's surface my beloved Country, Saudi Arabia. But then our planet appeared to me as whole, a globe without divisions and wars; and I prayed Allah...*" said King Fahd's nephew Prince Sultan ben Salman ben Abdelaziz Al-Saud, one of the passenger astronauts of the American "Challenger" shuttle. "*Earth is our cradle, but we cannot remain in a cradle forever*" wrote at the turn of the century a Russian small-town schoolteacher who formulated the theoretical foundations of future space travel like Oberth and Goddard did respectively in Germany and USA, Konstantin E. Tsiolkovsky.

After long-term missions, it is hard to see how astronauts can ever be the same again. Many of them asserted that they had changed as a result of having been in space. Among the APOLLO astronauts, Collins had left NASA to become a missionary, Mitchell resigned to study extra-sensory perception, Cooper expressed his scientific interest in the study of Unidentified Flying Objects (like his Soviet colleagues Popovich and Grechko); and more recently, SKYLAB astronaut Pogue left the astronaut corps to join an evangelical group, while James Irwing led an archaeological expedition in search of Noah's lost ark on Mount Ararat in Turkey. All this just means that the "outer space effect" manifests itself whith common, positive patterns showing the development of a spiritual dimension in the subjects involved, with interests ranging from the frontiers of science to philosophy and religion. All this can only make these men better. They developed in their inner self what we might call "cosmic consciousness" -a new and anti-anthropocentric viewpoint whose general acceptance surely would solve many problems of today's uneasy world.

Hoping that his colonies may really be a wonderful place to live, Gerald K. O'Neill just anticipated our conclusions. Today it is more and more evident that the space experience is affecting positively young people all over the world, unifying different cultures in a unique, general feeling: what the late British science fiction author John Windham (who died in 1969, the year man landed on the moon) defined, as the title of one of his books, The Outward Urge towards space and other worlds.

Recent studies both in the Western as well as in the Eastern world seem to confirm such indications and the conceptual progressive philosophical and technological loss of importance of the planet Earth as a base for Homo Sapiens in young people. For them G.K. O'Neill's "high frontier" concept is quite accepted.

In 1992, 500 years after the discovery of America, the first permanently inhabited space station of the United States would have been placed in orbit. Unfortunately, the "Challenger" tragedy will make this almost impossible, and the "Columbus" space station won't orbit around Earth before 1993.

But in Italy this important step may seem far behind, as a project in itself, to the Florentine architect Daniele Bedini, who designed in 1982 a large space station named "O.L.G.A. TOWN", standing for Organical, Linear, Geosynchronous, Advanced Town. Its unique feature is that, for the very first time, enphasis has been placed on architecture in space research. For this purpose he was asssited by an interdisciplinary team of Italian and American space scientists (Rome and Berkeley Universities and NASA Ames Research Center) and private consultants, includind myself as far as social sciences are concerned.

Important problems were solved, such as the moment in which the Earth's shadow intercepts the space station, the ratio between the station's dimensions, the speed of rotation around its axis, and the level of gravity that must be articicially recreated. Another important solution was the screen which would protect OLGA TOWN's inhabitants from harmful cosmic rays. Due consideration was given to the processing of the materials suitable for the construction of the station. They included new systems of waste-recycling, new methods of crop-growing and miscellaneous items such as clothing, furniture, transportation and so on.

OLGA TOWN is a huge and spindle-shaped third-generation space station about 3 miles long and more than 1 mile in diameter, designed for a geosynchronous orbit around the Earth, and is located approximately 36,000 kms from its surface. Additional twin modules of the same shape and size may be attached to the original OLGA TOWN structure, whenever the need arises for more room. In so doing, one could conceivably produce a Saturn-like "ring" of modules surrounding Earth for a total theoretical capacity of 40 billions inhabitants. The OLGA TOWN structure may accomodate 1,000,000 people and is made up of an aluminum alloy while the habitat is covered by a protective shield in a titanium alloy which protects the dwellers from the previously mentioned cosmic rays as well as meteorites and the thermic changes. There will be night and day on OLGA TOWN. Special optic fibres will allow light to enter as needed. In turn, whenever you look up, you will have the impression of seeing clear blue sky. For a night-time effect, all one has to do is reduce the intensity of the incoming light. Inside temperature will be a constant 23°C, producing a permanent spring-like effect increasing agricultural yeld as well (four harvests per year are expected).

OLGA TOWN can be reached by space shuttle, of course. The first part of the journey takes you 300 kms from Earth. The rest of the distance will be covered in a special "elevator" and three hours later one reaches the station.

Compared to other NASA projects of space stations for the XXIst century, this Italian project achieved a great saving of atmosphere, which instead only fills the living areas: that is, all the transparent domes located on the inner surface of OLGA TOWN's cylindrical structure. Inside the cylinder there are many domes within which everyday life takes place. In the center of each cylinder there is a huge sky-scraper-like structure. This is OLGA TOWN's core of activity: schools, offices, hospitals, department stores, movies, theatres, etc. A regular downtown hustle and bustle. Off to the sides, one finds parks, playgrounds, swimming pools. Mind you, real parks with real trees, real grass, and real flowers. Another advantage offered by the domes is that you won't see other people above you upside down, like in G.K.O'Neill's future space colonies pioneer schemes whose unpleasant psychological effects won't affect OLGA Towners. NASA has already acknowledged the potential of this project and it may well be used as a guide-line for the (first-generation) 1992/1993 space station. In fact Bedini's studies showed unexpected practical benefits from the standpoint of both economy and habitability.

Technical and scientific know-how, history and philosophy have been combined to produce the essential features of OLGA TOWN. In so doing, the needs of the man of tomorrow have been envisioned. This new "Homo Sapiens" who will inhabit space will inevitably have a different outlook of life and the universe. From the psycho-sociological and ethical point of view, progress will probably make him better. OLGA TOWN has been conceived for this new kind of man: "Homo Cosmicus". And sure it achieves special significance which extends beyond mere technology and design: it becomes a symbol of mankind's hopes for the future.

A basic objective of the project was to make OLGA TOWN a place where life could be qualitatively better than on Earth and also more fulfilling. From an architectural point of view, OLGA TOWN's living areas inside the domes are a reflection of the old 19th century large steel and glass structures (more specifically, greenhouse-like) where one has the sensation of spaciousness and not of suffocation.

The houses are on three floors. The ground floor is designed for activity, work, entertaining friends: the second floor is for relaxation and sleep while the third floor will be equipped with a solarium, gymnasium and hot tub. Altough the emphasis will be on community living, OLGA TOWNners are also free to be social or anti-social, to live in togetherness or in isolation as any person sees fit.

To save much more room all the collective facilities, farmlands and animal breeding areas are not located on the same level as in other NASA projects, but on lower levels beneath the domes themselves. Greenhouses are set up for the cultivation of vegetables, above all soy, and plants for the zootechnical breeding of fish, rabbits and small animals. Furthermore, they contain all the recycling processes. The biological cycle in space is complete, including the chlorophyll-photosynthesis one. This is made possible because natural light reaches the plants through optic fibres.

The residential domes have two heights: 25 meters for the living areas and 50 meters for the green areas. The two different heights give one the same psychological sensation of leaving "home" and being outdoors: a very important factor for a sense of well being for people isolated in outer space.

OLGA TOWN's most original feature is that it is an architectural project. Up to 1982 no one had ever really given due consideration to space architecture. Bedini had. And this before NASA included it in its space programs and in the planning of its studies. His project deals with everything, from interior decoration to the outer structures, to landascaping. Nothing is left to chance.

So, if OLGA TOWN were to be built now, as he had planned it, one could move in immediately and feel perfectly comfortable and relaxed. OLGA TOWN is full of colors: it is and must be lively. Life there will be easy-going and relaxed. OLGA TOWNers will have the utmost degree of freedom. They will have all the time in the "world" for their intellectual betterment and for their own personal fulfillment. In brief, they will enjoy the best things that Earth has given them while benefiting from what the future has in store for them.

G.K. O'Neill's pioneer studies in the Seventies seem far away. The "high frontier" concept is more and more evolving. The future will turn it into tomorrow's facts.

References

[1] Asimov, I. & McCall, R., Our World in Space, New York 1974, (Italian Transl.: l'Uomo Nello Spazio, Milano, 1975)

[2] Bedini, D, Beyond Columbus: Italy's Solution to Life in Space, in L5 News, May 1986

[3] David, W., Public Attitudes about Space as Surveyed by the National Commission on Space, IAF 86-358

[4] Florini, A., The Next Giant Leap in Space: An American Ctizen's Study of the Prospects for International Cooperation in Space, IAF 86-357

[5] Geis & Florin with Beren & Kelly, Ed., Worlds beyond/The Everlasting Frontier, Berkeley, Calif., 1978

[6] Glenn, J.C. & Robinson, G.S., Space Trek, New York, 1978

[7] Hawmey, T.B., Space Generation: Creating a Vehicle to Foster International Space Education and Collaboration, IAF 86-360

[8] Heppenheimer, T.A., Colonies in Space, New York, 1978

[9] Maruyama, M., Extraterrestrial Community Design: Psychological and Cultural Considerations, in Cybernetica (Belgium), 19, 1976

[10] National Commission on Space, Pioneering the Space Frontier: Our Next 50 Years in Space, New York, 1986

[11] O'Neill, G.K., The High Frontier, New York, 1977

[12] Powers, R.M., The Coattails of God, New York, 1981

[13] Serafimov, K.B., The Space Activity as a Bridge between the Young People and the High Technologies, IAF 86-359

Interventions

Lubos Perek

I think that the OLGA Town raises the question: why should it be located in the geostationary orbit? It is the most crowded part of outer space. For such a large station even the material crowding might be of importance; but quite critical would be the crowding in communications. Already now it is difficult to accommodate new space systems in the geostationary orbit.

Arnold H.W. Woodruff

I think it is important to raise the question how such a project as OLGA TOWN be financed?

Will it be paid for by taxpayers money? Whether taxes raised from the 1 million inhabitants or from all members of the owner nation or group of nations (i.e. central government funds); or will it somehow be commercially financed -whether for example as an off-world holiday centre, or as accommodation for a consortium of companies wishing to staff space factories, or just as alternative (very expensive and only for the rich) housing?

Surely pure scientific curiosity will never be a sufficient condition, even if it is a necessary condition, to build such a vast high cost structure. To justify the enormous levels of finance necessary I feel another purpose to the benefit of mankind will be needed (whether central government or commercially financed). Such justifications could include the commercial ventures I've mentioned, or perhaps the large scale accommodation for major resource exploitation in a hostile environment (e.g. in Venus orbit) or mildly possibly for military purposes.

LA CONQUETE SPATIALE, CONDENSATION DU "SONGE" PHILOSOPHIQUE KEPLERIEN

"Que nous reste-t-il donc,
si ce n'est de dire, avec Platon,
θεὸν ἀεὶ γεωμετρεῖν"
Kepler,
Mysterium Cosmographicum (1595-1596)

"de telle sorte que, chez eux (les lunaires),
soit entièrement autre la raison d'astronomie"
Kepler, Somnium (1609-1634)

Jean-Pierre Faye
Président de l'Université philosophique européenne
Université européenne de la Recherche
1, rue Descartes, 75005 Paris, France

Nous sommes à l'aube d'une ère de la connaissance de l'univers, comparable à celle qu'a ouverte la révolution scientifique de l'âge classique.

Pour reprendre les termes du Directeur des Programmes Scientifiques de l'Agence Spatiale Européenne dans une publication récente [1], *"avec l'utilisation des techniques spatiales, l'astronomie va sous peu connaître un développement aussi révolutionnaire que celui qui suivit au XVIIe siècle l'invention du télescope"*.

1. Un espace de phase : de la Narratio Prima au Somnium

Je voudrais tenter de faire apparaître quelques uns des fils rouges qui relient le développement révolutionnaire de l'astronomie spatiale, demain, à la révolution scientifique, hier, -celle de l'âge classique, qu'ouvre le renversement copernicien, l'<u>Umwälzung</u>, comme l'appelle Emmanuel Kant : cette <u>phase</u> qui s'ouvre avec la <u>Narratio Prima</u> de Rheticus et se poursuit jusqu'à l'<u>Astronomia nova</u> de Kepler et les <u>Dialoghi</u> de Galileo Galilei...

... pour atteindre toute son amplitude avec la <u>Meditatio nova</u> de Leibniz et les <u>Philosophiae naturalis principia</u> de Newton, dont nous célébrons, cette année, la fête jointe. Tricentenaire, certes. Fête européenne et mondiale de la pensée.

Il devrait être possible d'explorer cet "espace de phase", de telle façon que sa <u>Narratio</u> et sa <u>Meditatio</u>, son Astronomia et sa <u>Philosophia naturalis</u> laissent entrevoir quelque secret de son énergie.

J. Schneider and M. Léger-Orine (eds.), Frontiers and Space Conquest, 59–67.
© 1988 by Kluwer Academic Publishers.

Au coeur de cet espace de temps, déployé, disons, de 1540 à 1687, un ouvrage singulier se détache, à mi-temps (ou au mitan), un livre fait de dix pages de texte ou de récit, et de quarante pages de Notes -de 223 Notes-, plus un Appendice d'une page et dix pages de Notes sur cet Appendice.

Ce livre, écrit entre 1609 et 1623, mais paru après la mort de Kepler, en 1634, s'intitule alors :

Somnium
seu

Opus Posthumum

de Astronomia
Lunari

Il paraît à Sagan, en Silésie. C'est le Songe d'une pensée scientifique rigoureuse, la première sans doute à introduire des lois exactes dans l'univers physique -mais c'est le phantasme singulier d'un songeur, qui va se révéler, pour lui et sa famille, étrangement "funeste" et "omineux" -ominosum. Car on a fabulé, ou "confabulé" sur sa fable, jusque "dans les échoppes de barbier". Et l'effet en fut un procès en sorcellerie de six années, au cours duquel sa propre mère subit quatorze mois de prison -procès qui s'achève en 1621 et qui résonne comme le prélude d'un autre procès, celui de 1633.

LE "SONGE" COMME CONDENSATION

Or ce songe est comme le frontispice, antidaté de trois siècles et demi, de l'astronomie spatiale d'aujourd'hui et de demain, ou même de l'astronomie in situ. Celle-ci est à son tour comme la réalisation condensée ou la condensation réalisée de sa fantasmatique, -de même que ce rêve est comme la Verdichtung, la condensation onirique de tout le débat philosophique de l'antiquité, de la Renaissance et de l'âge classique. Et même, par certains traits, de la méditation relativiste de notre siècle, point de départ de la seconde "révolution astronomique".

Le but de mon Songe, dit expressément Kepler, est de donner un argument en faveur du mouvement de la Terre. Ou plutôt, d'utiliser l'exemple de la Lune comme "*solution*" aux objections venues, selon son expression curieuse, "*de la contradiction universelle du genre humain*" -ab universali contradictione gentis humanae. On connaît l'épreuve ultérieure de cette contradiction. Luther en 1539 a traité de fou -de Narr- "*l'astrologue nouveau qui voulait que la Terre se meuve et non le Ciel*". Et si Alexandre Farnèse -le pape Paul III- n'a pas sourcillé en recevant la dédicace de Copernic, Matteo Barberini -Urbain VIII- sera le pape du procès de Galilée.

FICTION RELATIVISTE

Mais la contradiction de la gent humaine s'explore ici d'une autre façon et trouve sa "solution" autrement, -en des termes "toujours futurs" et pour ainsi dire prospectifs.

Non seulement parce que Kepler donne argument aux hypothèses de Copernic. Mais aussi lorsque, dans la Note 66 du Songe par exemple, il définit la gravité "*comme force d'attraction mutuelle*" [2]. Copernicien, il est déjà prénewtonien, par les termes, comme par les lois dont Newton va donner la démonstration. Mais la fabula du Songe va rendre possible un décentrement plus intrigant : car la montée dans l'ile de Levania -la Lune- permet de supposer "*qu'un observateur situé sur la Lune penserait qu'elle reste absolument immobile*", ..."*qui in Luna esset, omnino in Lunam stare loco fixam existimaret*" [3]. Ce qui in Luna esset, que Michèle Ducos a tout naturellement, et fort pertinemment, traduit par "un observateur situé sur", introduit un effet de narration qui dépasse alors les possibilités de démonstration et par lui, un décentrement des référentiels, l'apparition d'un observateur au subjonctif, capable de percevoir le mouvement comme fixe et l'immobile comme en mouvement. Car il était nécessaire ici de "*parler selon l'imagination de la vue*" -hic ad imaginationem visus loquendum. Ainsi "*pour ses habitants Levania est non moins immobile au milieu des astres que la Terre pour nous les hommes*". Ce nec minus immota, n'est pas seulement un "argument de copernicien". Déjà la fiction d'une astronomie spatiale a fait pénétrer d'avance la pensée dans la physique d'un univers relativiste. C'est Levania, la Lune qui est immobile. C'est la Terre, Volva, qui tourne.

Cette interrelation entre la fiction et la conquête spatiale -et leur effet en retour sur la pensée théorique-, voilà ce qui est peut-être commun (mais c'est une question que je pose) à ce moment singulier du Somnium de 1609, comme Astronomia Lunaris, et à l'actuel "développement révolutionnaire" de l'astronomie spatiale.

2. Le "choc initial" du voyage spatial

Approchons donc du moment initial : le vol spatial vers Levania, la lumineuse. Il est placé sous les signes de la magie, non de la mécanique, et attribué au "Daemon ex Levania". Mais celui-ci est défini par la Note 72 : "*le démon que l'on appelle astronomie...*" Et voici le récit démonique : "*tous ensemble (...) nous le poussons par dessous et l'élevons ainsi dans les airs...*" -subtus nitentes in altum eum tollimus. Alors, poursuit le Songe, "*le choc initial est très pénible pour lui*" : Prima quaeque molitio durissima ipsi accidit. Cette Prima molitio, ce déplacement initial fait entrer en jeu le concept de masse -moles- qui va se construire précisément par les lois de Kepler et la physique galiléenne : il trouvera dans le principe newtonien de l'égalité entre action et réaction, qui en procède, le ressort même du futur moteur spatial. Molitio durissima, en effet, choc très dur du déplacement initial d'une masse dans l'espace, quand elle s'arrache à l'attraction keplerienne et newtonienne, dans le cas, supposé par la Note 60, "*d'un corps violemment arraché à la Terre*" et "*transporté sur la Lune*" : "...erit forte et hoc possibile ut corpus aliquod Terris divulsum importetur in Lunam". L'importation dans la Lune d'un divulsum, d'un mouvement arraché à la gravité : nous voici dans la fiction pure. Et déjà, avec trois siècles et demi d'avance, dans la conquête spatiale.

LES CONDITIONS FRONTIERES

Mais l'imaginaire Keplerien s'attache en même temps à définir ce que l'on pourrait appeler les conditions frontières, ou les conditions d'empêchement.

Ainsi sur Privolva, face cachée de Levania, ou plutôt hémisphère lunaire qui est "privé de la vue de Volva", la Terre -la période diurne d'exposition au Soleil dure quatorze de nos jours, aussi la chaleur y est extrême, et l'eau en surface, chauffée par le Soleil, est bouillante... Il s'ensuit que "*les plongeurs sont très nombreux*", les êtres animés "*vivent sous les eaux dans la profondeur, et leur technique -leur art- vient en aide à la nature*"...

Je souhaiterais entendre cette description en confrontation avec celle-ci : "*Sans les moyens spatiaux, l'astronome pourrait facilement comparer sa situation à celle d'un crustacé vivant au fond de l'océan et qui chercherait à découvrir la source de la lueur qu'il aperçoit vers le haut. Il peut, bien sûr, conjecturer l'existence de cette source et les causes de son déplacement apparent. Il sait aussi que la masse liquide qui le sépare de cette source est la cause de son incapacité à en savoir davantage sur elle. Mais imaginons qu'il invente une machine qui lui permette de nager jusqu'à la surface : l'horizon de ses connaissances change alors radicalement lorsqu'il découvre les nuages, la Lune, le Soleil, les étoiles...*" [1]

RAPPORT A L'OBSERVATEUR : LES CHEMINS MOUVANTS

La pensée des conditions frontières appartient en effet à l'astronomie spatiale, et à la conquête spatiale elle-même. L'interroger, et en décrire les champs, c'est explorer le rapport du sujet observant à son univers. C'est en repensant ce rapport que la pensée cosmologique et la connaissance de l'univers sont allées en gagnant leur "profondeur", précisément : leur profondeur de champ.

C'est pourquoi il importe de prendre une vue serrée de ces moments de renversement, ou de retournement des chemins. Et des versions approchées qui en sont données. Y compris celle de Martianus Capella (Minneius Felix) dans son Satiricon [4] rapportant la vue selon laquelle Venus et Mercure ont le centre de leurs mouvements dans le Soleil [5], version que rappelle Copernic dans le De Revolutionibus -ce même Capella qui se trouve être le premier chez qui apparaît le terme de sujet : subjectum.

Ou bien celle de l'énigmatique auteur du Cinquième livre [6] de Pantagruel, qui est sans doute (mais non certainement) ce François

Rabelais dont je rappellerai la présence comme étudiant dans les lieux où nous sommes -l'ex-Collège de Navarre. Lorsqu'il décrit "*les allures de ces chemins nouvaux*", dont Pantagruel nous dit "*que (...) Philolaus, Aristarchus avoient en icelle isle philosophé, Seleucus prit opinion d'affirmer la terre veritablement autour des poles se mouvoir, non le Ciel encore qu'il nous semble le contraire être vérité ; comme, étant sur la rivière Loire, nous semblent les arbres prochains se mouvoir, toutefois ils ne se meuvent, mais nous par le decours du batteau*". Ou encore ces mots de Plutarque que transcrit Copernic lui-même : "*D'autres pensent que la Terre se meut ; ainsi Philolaus le Pythagoricien dit qu'elle se meut autour du feu dans un cercle oblique...*"

L'ILE UNIVERS

L'Isle d'Odes du Cinquième Livre, cette Isle des chemins -ὁδός- est celle où "*les chemins cheminent comme animaux et sont les uns chemins errans, à la semblance des planètes ; autres chemins passans, chemins croisans, chemins traversans. Et vy que les voyageurs (...) demandoient : "Où va ce chemin ?"*".

Cette Ile ressemble aux univers-îles, d'un monde où les référentiels des mouvements sont eux-mêmes en déplacements. Le double moment théorique de 1917 et 1924 -quand, aussitôt après la Relativité générale, Hubble démontre que les "nébuleuses" (ou ce que l'on nommait ainsi) sont "extragalactiques", c'est-à-dire sont elles-mêmes des galaxies- est déjà présent dans le moment qui va de 1540 jusqu'à 1610 : date du Somnium, de la découverte des satellites de Jupiter et bientôt de la Nébuleuse d'Andromède (1612). Kepler, Galilée et Simon Marius à cet égard sont les contemporains de Hubble...

Il nous faut être attentifs à la pensée même des frontières de la conquête spatiale, sous ses deux formes -approfondissement optique et déplacement matériel, permettant à son tour un nouvel élargissement (ainsi le Télescope Spatial sera un multiplicateur 50, par rapport à Palomar ou Zelentchuk). Car c'est à cette pensée des frontières que prend appui un nouveau renversement des perspectives. C'est l'idée que la Via Lactea, la Galaxia était la frontière de notre "habitation" d'univers, contenant dans ses bornes l'habitant Soleil, qui a suscité le rebondissement d'une nouvelle échelle cosmologique.

3. Prouver la simultanéité

Entre le moment Kepler/Simon Marius et le moment Einstein/Hubble, il y a un opérateur de pensée, qui est précisément le penseur des limites de la pensée. C'est le moment kantien : c'est celui de la Théorie du Ciel de 1755 -où il s'agit déjà de penser que le système solaire est "comme" une Nébuleuse d'Andromède- ; mais c'est aussi celui de ce paragraphe de la Critique de la Raison pure contenant la petite phrase qui aurait déclenché la découverte théorique de la Relativité [7] : "*la lumière, qui joue entre notre oeil et les corps d'univers, produit une communauté médiate entre eux et nous, et par là en prouve la simultanéité" "...Das Licht, welches zwischen unserem Auge und den Weltkörpern spielt, eine mittelbare Gemeinschaft zwischen uns und diesen bewirkt und dadurch das Zugleichsein der letzteren beweist"* [8]. Que dans un écrit de 1781 et 1787 soit soulignée cette fonction de la lumière, comme ce jeu qui produit la "communauté médiate" entre les corps d'univers et l'oeil de l'observateur-sujet, et ainsi le fasse témoigner de la simultanéité, ou de l'être-en-même-temps (das Zugleichsein beweisen), voilà qui est en effet un tournant décisif de la pensée philosophique engagée aux frontières de la pratique physicienne.

LA FIGURE INASSIGNABLE ET LE SUJET

Il est important d'accentuer le fait que la conception intialement la plus marquée des frontières d'univers -et en même temps de la relation des mesures d'univers à la lumière et aux observateurs qui en "témoignent", qui en font le Beweis -se trouve liée à la philosophie du sujet, du denkend Subjekt : la philosophie kantienne. A mi-chemin entre Kant et Kepler, à son tour, le moment newtonien n'est pas pleinement pensable sans son double leibnizien : celui de la Meditatio Nova de 1686 et de "l'analyse des infinis". Et la relation pensive entre la découverte du "triangle inassignable" [9], sur la figure de Pascal, et la conception d'un sujet qui va s'évanouissant -d'une sorte d'état quantique de la conscience percevant, "d'homme microscopique", au sens de Pierre Auger- en est la clé.

LE QUADRIVECTEUR, LE PRODIGE, L'EVENEMENT

L'apprentissage imaginaire de la conquête spatiale a donc son commencement -et son secret sommet sans doute- dans le Somnium keplerien. Et il est l'éducation au changement des systèmes de référence : à cet apprentissage du change. Et il n'est pas tout à fait curprenant que l'exercice libre du pouvoir de référer, c'est-à-dire de narrer, de conter -du referto en langue italienne, du referente en langue espagnole-, précède, prépare, accompagne les grands renversements de la pensée scientifique.

C'est dans l' ἱστορίη , livre II, chapitre 32 d'Hérodote -narration, information, enquête, connaissance de l'ἵστωρ, le Sachant, le Narrator, le Gnarus, le Gnoscens-, qu'apparaît la première définition écrite de ce quadrivecteur (x,y,z,ct) qu'est l'événement : "*Ces Egyptiens, lorsque survient un prodige, le mettent par écrit, et observent de quel événement il sera suivi. Si, dans la suite, il arrive quelque chose qui ait avec ce prodige la moindre ressemblance, ils se persuadent que l'issue sera la même*" [10]. Par rapport à l'observateur terrestre, est prodige tout événement de l'astronomie, à commencer par le "pas gigantesque" du 21 Juillet 1969, selon les termes de Neil Armstrong : nouveau Pantagruel. Mais avec le Télescope Spatial, quand il sera en orbite, c'est la définition même de "l'événement" par excellence, la question de la "singularité initiale", qui sera sans doute approchée autrement -à travers l'exploration plus "éloignée" du red shift.... Ainsi, déjà, de l'observation des quasars par le satellite Einstein depuis 1978 dans le domaine de l'astronomie X, dont la sensibilité à l'expérience a été, avec lui, multipliée par mille.

CONQUETE SPATIALE ET NARRATION

C'est en lisant La Guerre des mondes de Wells que Robert Hutchings Goddard décide de consacrer sa vie à la conquête du voyage spatial, pour aboutir au lancement, dès le 16 Mars 1926, de la première fusée à propergol liquide. Or avec Les premiers Hommes dans la Lune de H.G. Wells, il s'agit d'un récit fortement inspiré par le Songe de Kepler : dans les deux cas, par exemple, la végétation profite de "*la brève poussée du jour lunaire*", pendant laquelle elle doit "*fleurir, fructifier, se semer et mourir*"...

Ainsi le pêcheur insulaire chez qui se réfugie Duracotus, le narrateur et voyageur spatial du Somnium, porteur d'une lettre à Tycho Brahé, -ce pêcheur de l'île de Hveen est un "véritable philosophe", verus philosophicus, qui a pour habitude de perpétuellement questionner -percontari-, et de "*faire grand cas des narrations de cette sorte*" : magni facere narrationes hujus modi.

Au coeur du voyage spatial, et de l'astronomie spatiale, cette conquête permanente ou cette cueillette du "prodige", agit ce processus transparent (au point d'échapper presque à notre surprise), mais qui est lui-même pourtant une énigme curieuse au sein de l'univers : la narration, qui rapporte au sujet, à l'observateur et qui éventuellement interroge et demande à rendre compte, à lui, -narrateur, observateur, "perconteur".

Sur le lieu d'où nous parlons, qui fut la rampe de lancement de cette fort curieuse entité européenne, l'universitas, le moment vient en effet de songer de façon conséquente à cette condensation du savoir, à cette micro-Université qu'est un module ou un "mode" d'astronomie spatiale. Porteuse de tous les savoirs, depuis la physique quantique jusqu'à l'anthropologie du futur, de l'astronomie des quasars jusqu'à la microphysique. Et porteuse aussi de la narration humaine : narration nietzschéenne de la puissance, -et "biographie interne" ou "mémoires intérieurs" de la fragilité.

PUISSANCE ET FRAGILITE

La narration de la puissance spatiale passe par la figure, pantagruélique, ou hypergargantuine, du géant de Poincaré : "*aux yeux d'un géant pour qui nos Soleils seraient pour nous nos atomes, la Voie Lactée ne semblerait qu'une bulle de gaz*" [11].

La narration de la fragilité anthropologique passe par la Seconde partie des Voyages de Gulliver. Voyage à Brobdingnag. Séjour dans la "boîte fermée de toutes parts", tirée par des monstres.

Nous aurons désormais toujours davantage à penser par les frontières de ces narrations de la puissance et de la fragilité.

Le voyage spatial est alors le prolongement corporel de cette maîtrise dans la puissance de rapporter, dont Kant donne la description admirable dans la Théorie du Ciel. Le traité de Thomas Wright l'a conduit, dit-il, à "*considérer les étoiles fixes non pas comme un fourmillement dispersé sans ordre visible, mais comme un système qui a la plus grande ressemblance avec un système de planètes, de telle sorte que, comme dans celui-ci les planètes se trouvent très près d'un plan commun, de même les positions des étoiles fixes se rapportent d'aussi près que possible à un certain plan qu'il faut étendre par la pensée à travers tout le ciel*". Le déplacement du rapporter va être un changement de puissance : "*par leur amoncellement plus dense sur ce plan, elles (les étoiles) représentent cette bande lumineuse que l'on nomme Voie Lactée (...) j'ai trouvé que (...) les étoiles dites fixes (...) pourraient bien être proprement des étoiles errantes au mouvement (...) d'un ordre plus élevé*".

La conquête spatiale poursuit cette torsion narrative que la pensée introduit dans le "fourmillement dispersé". L'arrachage au sol y prend son principe. La prima molitio est l'effet, rêvé et réalisé, de la Narratio prima. L'approche toujours plus attentive de leur jointure est le problème qui s'ajoute sans cesse à tous les autres.

Le voyage spatial, c'est en effet ce plan étendu à travers tout le ciel par la pensée, qui le rapporte à des références nouvelles, et qui ainsi se fait captatrice de nouvelles énergies.

Références

[1] Roger Bonnet, in : Astronomie Flammarion, 1985, T.I., p.206
[2] Attractio mutua
[3] Note 54
[4] De Nuptis Phlologicae et Mercuri (vers 470)
[5] C'est la théorie dite "égyptienne" d'Héraclide de Pont
[6] Paru en 1564 (L'Isle Sonnante en 1562). S'il est bien de Rabelais, ce qui est assez probable, il a donc été écrit avant 1553, soit dans la décennie du De Revolutionibus.
[7] Selon le témoignage d'Albert Rivaud, qui fut secrétaire d'Henri Poincaré, Einstein aurait signalé l'importance, pour lui, de cette notation de la Kritik kantienne.
[8] Kritik der reinen Vernunft, Hamburg 1952, p. 262
[9] Homothétique du triangle des coordonnées cartésiennes
[10] Noté par Jean Rösch, in Astronomie Flammarion, 1985, T. I, p. 24
[11] Science et méthode

Intervention

Maurice de Gandillac

La source de Kepler est probablement N. de Cuse (peut-être par l'intermédiaire de Bruno). Mais l'essentiel est l'infinité du monde et l'absence de tout point fixe.

D'autre part, Kant était aux antipodes de la relativité einsteinienne (le temps et l'espace sont des formes a priori).

THE ART OF EXTRAPOLATION

Henk C. Van de Hulst
Leiden observatory
Sterrewacht Leiden
P.O. Box 9513, 2300 RA Leiden, The Netherlands

1. Introduction

This will be a lecture about space and about man. That much was clear from the start. It also has to be a lecture about myself. The necessity of this personal touch came to me as a surprise and did not make the preparation easier. Let me briefly tell you how it went.

The first invitation, which reached me by mail last summer, invoked a clear and immediate reaction. I did not understand a single word of the title suggested for my talk, so I wrote back: "This is crazy enough to be a great challenge ! Yes, I accept, but please tell me what you mean."

After a long talk with Jean-claude Pecker in November I felt inspired to prepare an objective, all-inclusive review of the non-technical aspects of man in space, I deliberately say "non-technical" rather than "philosophical", "psychological" or "cultural" since these words already have a tendency of narrowing down the subject. By saying "objective and all-inclusive" I mean that the type of lecture I prefer to give is about a subject I <u>know</u>. There are some subjects about which I could give such a talk, e.g. radiative transfer. Typical for such a talk is that the speaker has gone over the subject so well that his personal contributions and discoveries have been polished out (although some personal likes and dislikes may still shine through). He can then literally stands aside as a person and say: "Look at this beautifully coherent picture and enjoy it while I point out some details which may help you to deepen your appreciation".

Alas, this was not the final stage. The more I tried, the more I realized that giving an objective lecture was a prepostorous idea. I simply do not know enough to treat the subject man-in-space objectively. I know a good deal about space projects, but by no means everything, and certainly not about the more far reaching manned space projects of fantasies. I also know a good deal about man and his aspirations but this is a negligible fraction of what there is to know. Through many years I have made random visits into history, poetry, philosophy, psychology and religion. In doing so I have explored many paths, some because they were recommended to me and some because they seemed exciting. But these paths do not <u>cover</u> the

J. Schneider and M. Léger-Orine (eds.), Frontiers and Space Conquest, 69–86.
© 1988 by Kluwer Academic Publishers.

field. Instead they remain narrow pathways meandering through a largely unknown territory. To put it more as an adventure: I may regard a path as familiar, after treading it a dozen times, but still not know that just to the right there is a dangerous swamp or just to the left there is a broader or easier road.

Listening in this frame of mind to yesterday's philosophical introductions with the dozen familiar names gave me a slight feeling of missing something. Unwittingly, I had hoped for something with the quality of a lightning bolt, or of morning dew. Many of you may recognize these attitudes.

Clearly, an objective treatment is impossible. Hence I decided to drop the restraints and to be frankly subjective, to the point of anecdotic. So permit me after this introduction to be your unauthorized guide and to show you a few places I have visited, pointing at some details I have felt happy or uneasy with.

In summary I shall deal with the following subjects.

- I start with **Extrapolation in Science**, a somewhat technical subject, which many of you, who are scientists yourselves, will be able to recognize. I show by some examples that extrapolation is always insecure but can be made rather firm by using an asymptotic theory.

- **Utopias** is the subject of the next section. Here the extrapolation is in time and the subject is society. The extrapolation in time is called a prophecy and an asymptotic theory for society is called an utopia. I make some comments on utopias generally and more in particular about the utopias on space exploration and space habitation that have flourished most openly in the early 1960s but are still quite alive among enthousiasts, statesmen and participants of this symposium.

- **Management**, the subject of the next section, brings me closer to the present time. In any planning, say, for the next 5 years, it is necessary to decide what the next move should be and then to make that move. This extrapolation to the near future has its own hazards and I shall comment on some of them.

- The final section is called **Moods**, which is a short word for the complex of psychological factors that are relevant to the life of persons or groups in space. In planning an instrument we are used to look carefully at the various modes in which it may function, some designed for use, some usually called malfunctioning. In strict analogy we have to examine the various moods in which a person or group may function, including both the intended and the unintended moods. Since this subject is farthest from my own field I can most freely speculate on it.

2. Extrapolation in science [1]

Since many of you are natural scientists of one kind or another, I presume that a few simple examples from science, in particular from astrophysics, are acceptable. After all, we are here in the rue Descartes and all of us, scientists and philosophers alike, live in a world permeated by Cartesian coordinate frames. Therefore, we shall draw one figure (figure 1), where the horizontal coordinate is any variable (for which we later shall take time) and the vertical coordinate is a quantity which depends on the value of that variable. Suppose part of the curve portraying that dependence is known from measurements or direct observation. The art of extrapolation is to guess (or to compute) how this curve continues.

Figure 2 shows a mechanical analogue. The part of the broad which is in the clamp is the known, fully fixed part. At some distance from the clamp the board may bend up and down. This will render practical work at that position next to impossible, as you soon find out if you try to hammer in a nail. The classical remedy is to provide for some extra support, either at the work spot or at some further distance as drawn. This extra support is the metaphor for an asymptotic theory, i.e. a statement based on theoretical insight or calculation about the behaviour of the curve at large values of the variable, up to infinity.

Now turn again to figure 1. A mathematician, unacquainted with the physics of the problem, might conjecture that this is part of an oscillatory curve (b) and he might estimate both the amplitude and the period of this oscillation. Another one might guess that it is a curve gradually reaching saturation (curve a). But an astrophysicist, who knows that the abscissa is the number of absorbing atoms of a certain species and the ordinate the strength of the absorption line produced by that species in a stellar atmosphere, knows (most students have studied this subject in class) that the likely extrapolation is curve d, which looks rather incredible but that is the way nature has arranged it. Absorption lines in the atmosphere of the earth yield an extrapolation more like curve c. The moral of these examples is: guessing is insecure; only a good asymptotic theory will lend enough support for a reliable extrapolation.

In rare cases the asymptotic theory is so simple and clear that any extrapolation can be said to be a mere interpolation between here and infinity. Figure 3 shows an example from my book on radiative transfer, where both the quantity plotted (never mind the exact meaning) and the horizontal scale have been chosen to fit the asymptotic theory. The effect is that the initially formidable task, to extrapolate beyond $b = 1$, has become the filling in of a very minor portion of an already well-determined curve. Counting 1, 2, 3, infinity, has become simple in this example.

Our actual subject (sec. 3) is reached from these examples by two changes. We shall first make only one change: the independent variable is now taken to be time, but we stay for a while in science. Extrapolation is then called prediction. Well known examples are predicting the motion of the planets and predicting the weather.

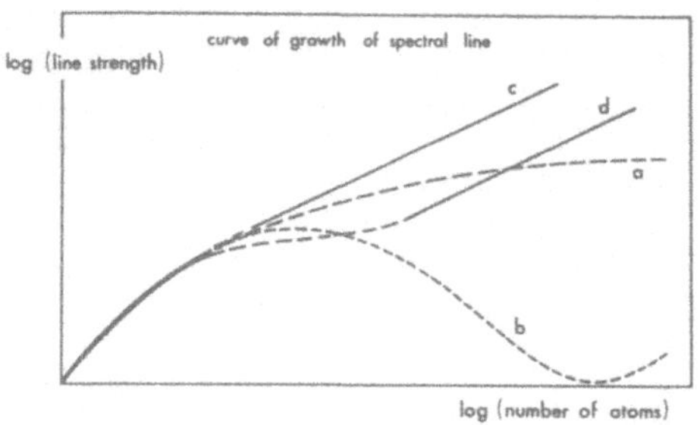

Figure n° 1
Curves of growth.
The same gently bending curve is extrapolated in 4 legitimite ways, using different asymptotic theories

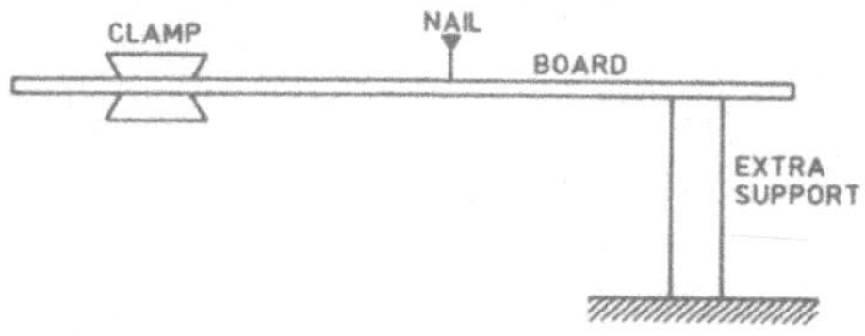

Figure n° 2
Extra support is carpenting.
In the metaphor the board is society, the clamp is the present, the nail is a future policy decision, the extra support is an asymptotic theory or utopia

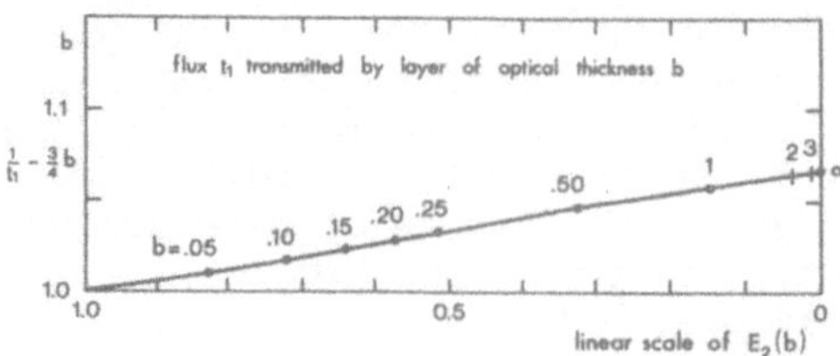

Figure n° 3
Extrapolation made into interpolation.
An example from radiative transfer

For a long time the hope has been to make weather prediction as reliable as predicting planet motion by using more complete data and ever bigger computers. The present trend is different. Fundamental barriers in penetrating far in time have been discovered even in fully deterministic dynamical problems [2]. This is a modern and advanced subject in mathematical physics and the study of chaos has become a specialism.

I have several times made a demonstration experiment in class by placing a bicycle upside down on top of the desk. Make sure that the front wheel has a little extra weight at one side and give it a fast and free spin. Plotting the angle covered against time elapsed gives a fast rising curve which gradually gets less steep because friction makes the wheel slow down. At the same time wiggles appear since the whell tends to rotate faster when the extra weight is down and slower when it is up (figure n° 4). At last the wheel stands nearly still when the weight is up and one wonders whether the weight will make it through the upper point or will fall back. At such a branch point the future course is inpredictable. Curves a and b in figure 4 are equally likely extrapolations. In each of these the wheel henceforth oscillates around the equilibrium position in which the weight is down. We say that lock-in has occured. The continuing friction gradually damps the amplitude of the oscillation. The phase in both curves is different.

Like the moon, the wheel henceforth has its same face turned to the earth. The oscillations of the moon are called librations. Only half of this example is a metaphor: the wheel is not the moon but the same earth pulls both. Similar lock-in phenomena and branch points occur in more complicated systems. If a third body is involved, a seemingly innocent phase difference as sketched may make all the difference in the motion of that third body and soon the entire system is inpredictable even though it remains fully deterministic.

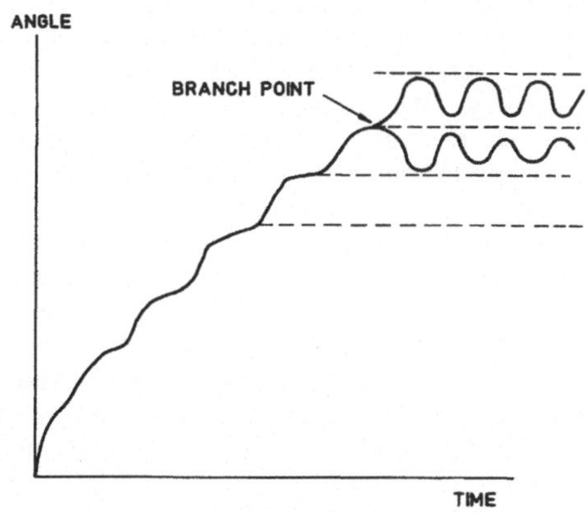

Figure n° 4
Lock-in. A class-room demonstration with a bicycle explaining how the moon came to show its same face to the earth

3. Utopias

We now make a second, major jump. Besides specifying that we deal with an extrapolation in time, we take human behaviour and society as the object of study. An extrapolation in time about such a momentous object is not simply called a prediction but a prophecy. And an asymptotic theory describing society in the distant future is called an utopia. There have been utopias as long as people have been together talking, singing songs or writing books. Many writers, painters, philosophers and religious leaders have depicted such utopias, often in earnest, sometimes in fancy with an element of wishful thinking.

In the usual connotation of the word the future depicted by utopia looks good. I shall extend the use of the word to include asymptotic theories predicting a bad future. In that respect Spengler's Decline of the West and the report Limits to Growth" of the Club of Rome were examples of utopias.

I do not have to explain the colour code going with that distinction. The usual utopias paint a rosy future but some see it black, or at least gloomy. Two further questions arise: to what extent is our idea about the future· coloured by our wishes, and to what extent may we permit that to be so? You should be aware that these questions receive widely different answers in different contexts. In science (as many of you know from a long practice) wishful thinking is bad and selection effects (i.e. grabbing facts and evidence that suit our wishes) are to be avoided. In religion and in mediative practices (at least in christianity and buddhism) wishful thinking does not have this bad tone and to wish ardently and consistently is positively recommended and sometimes trained. Many nuances and side remarks should be added but by and large, I hope, you see the contrast.

What is the colour of this conference? We shall see while we are here. But when I first saw the program, I had a funny experience. At one side I read that the meeting was to mark the threshold of the 21st century. On the other side, the explanations seemed to be coloured by a typically 19th-century belief in progress (more specifically: progress in society by the intelligent use of technical means). Of course, I realize that this idea of progress is by no means extinct and that it blossoms fully among an army of space enthusiasts. I cannot avoid to ask at this point whether personal wishes or promotional activities (of corporations or nations) have provided this heavy colouring. This question leads me away from my main topic, but it will pop up elsewhere.

Utopias do not only provide decorum, they have a real function. The practical function is quite analogous to the technical function of the extra support of a long slat of wood which I showed in figure 2. The extra support at the end provides just enough firmness to drive in a nail at the working place, which is much closer. You can find examples, for instance, in Marxist ideology.

If there is anything we have learned from practical space research, it is: never rely on an instrument before it has passed a number of stringent qualification tests. In the examples of figure 1, where time was not involved, the tests of the asymptotic theory and of the full curve leading to the asymptote were made by elaborate computations. But what qualification tests do we have for an utopia? Not even a shaking test would help for beliefs are not easily shaken. I suggest that, instead, we must turn to history and see how other utopias have fared. The record is interesting and to some extent amusing but does not leave us with great hopes that reliable predictions are possible.

I could do the further discussion in words only, but those of you who are used to reading diagrams will probably find a diagram more helpful. In figure 5 the abscissa is the time at which a statement is made and the ordinate the time about which the statement is made. The sign of the difference determines whether the statement is a historical observation or the prediction of a future event. A document like Horizon 2000 contains statements made in 1985 about, say 2005, which can be represented by a particular point in the diagram and the entire document is more like a vertical bar.

Having familiarized ourselves with the basic structure of this diagram, let us see what we can do with it.

In the past few years a number of nostalgic meetings have been held by national and international bodies (including ESA) celebrating 25 years in space activities. This meant a lot of reminiscing, represented in figure 5 by a cluster of points around A. We recalled how well we did in spite of the funny ideas we had in those early years after Sputnik 1. We also looked up, or remembered, some of the actual predictions we made at that time. Those early predictions are represented by a cluster of points about B, the mirror point of A with respect to the diagonal.

My personal record as a prophet is very poor. When in 1958 I became the first president of COSPAR, I saw enough in space research to devote a lot of my time to it. But I did not believe I would live to see man walk on the moon. I thought it would be too expensive. And seeing the way this line of research has been dropped afterwards I still wonder. I could cite dozens of other things I misjudged but probably many of us have the same experience.

Let me mention one more positive example [3]. An admirable paper written by Herb Friedman [4] in 1965, when X-ray astronomy was only 3 years old, predicts that by 1985 a new kind of astronomy would be flourishing along with radio and optical astronomy. He also predicted that the X-ray observations of "*hypothetical objects such as neutron stars*" would become important. These predictions have become true. He also points out "*prospects for the development of pictorial instruments utilizing emulsions or spark chambers and the electronics for rapid analysis of large numbers of events with the necessary selectivity*". This prediction was also correct. It is a surprisingly accurate characterization of the European COS-B gamma-ray satellite which so successfully flew 10-17 years later. But now listen: in the same paper Friedman foresaw that in 1985 "*life astronomers may feel quite at home on the moon*". He further speculated that X-ray rise and X-ray star set would enable the moon-based astronomers to obtain accurate positions with modest equipment and that the lunar base would house neutrino telescopes.

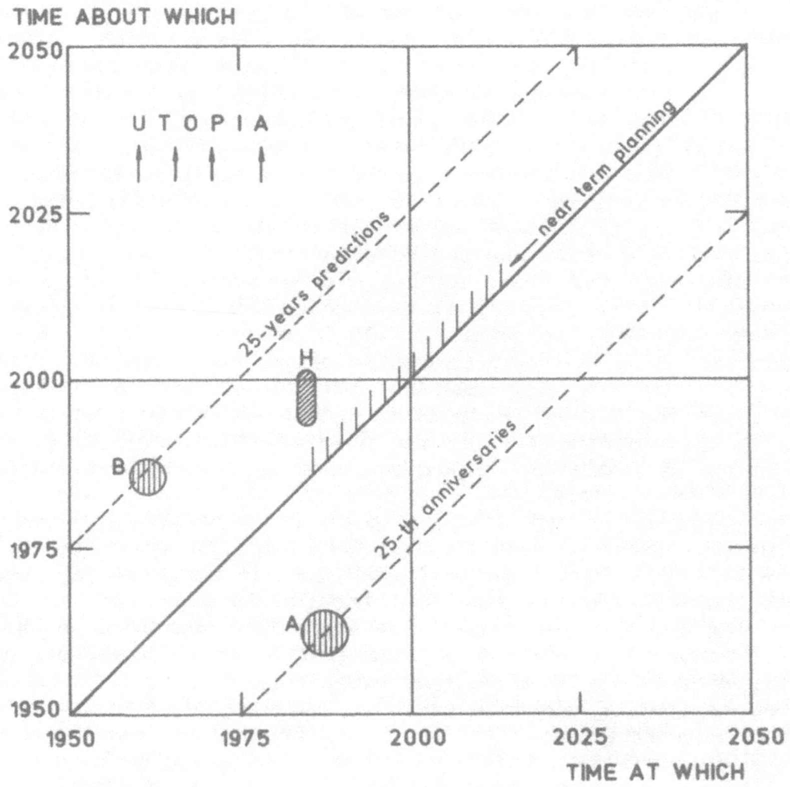

Figure n° 5
Predictions and historical statements.
Each point in the diagram stands for a statement made at a certain time (horizontal coordinate) about some other time (vertical coordinate)

The cautious conclusion is that predictions about milestones of manned space flight are far more difficult than those about unmanned space research. Or, they are so heavily loaded with wishful thinking that the term "prediction" hardly applies. Widely diverging opinions have existed from the early days and still persist. The first board of ESRO (or its predecessor COPERS), with Massey as president decided in the early sixties not to deal with man in space. In 1965 I had to review a book by Escobal [5] on Methods of Orbit Determination: Excercise 6 on page 237 of this book starts as follows: "*Geocentric Space Station Iota B-3 receives a distress call from the small space freighter Albatross*". My own feeling was: just how useful this book will be in finally saving the Albatross crew, remains difficult to judge.

I was equally sceptic but more deeply moved when I visited Minnaert in the hospital on one of the last days of his life. This was in 1970. Ha was a great rational idealist. He told me he was so happy that the IAU (International Astronomical Union) had finally agreed on the list of names given to new craters on the moon and that he was now working on a phonetic list to go with it. The committee members had taken the test to pronounce those names and they sounded rather different and (I quote Minnaert literally) "*That could be dangerous*".

While I still have this diagram (figure 5) on, I have a few further comments but, frankly, I am at a loss in what order to put them. Let me start with a simple one. Have you ever plotted an actual curve in such a diagram? I have, for I used to ask my students at their first arrival when they were planning to take their final exam and repeated that question many times later. Figure 6, which has in essence the same coordinates as figure 5, shows a typical curve, as well as the correct prediction curve. The typical curve is clearly optimistic. Should we therefore call the correct curve pessimistic? This would check with the definition of a pessimist which a friend in an industrial research lab once game me: "*a pessimist is someone who knows beforehand that it will take twice as long*".

A not so well known fact is the intrinsic structure which figure 5 has. If we show someone a description or a painting, carefully concealing both the year it was made and the year it is supposed to represent, which of these dates will be more readily guessed? the late Professor H. Van de Wall of the art department of Leiden University studied this question in the margin of a wider study on falsifications in art [6]. The answer was striking: no matter how much the artist has tried to make the picture look futuristic, he cannot drop the cultural signature of his time and a careful examiner may often date the origin to within a few years. This means that figure 5 has a kind of vertical striation. There was a wave of optimism about space travel in the early sixties pervading predictions then made for all future. In the same way there will be a rift at the date of the Challenger disaster. Predictions made in the post-challenger era will never be the same as they were before.

Figure n° 6

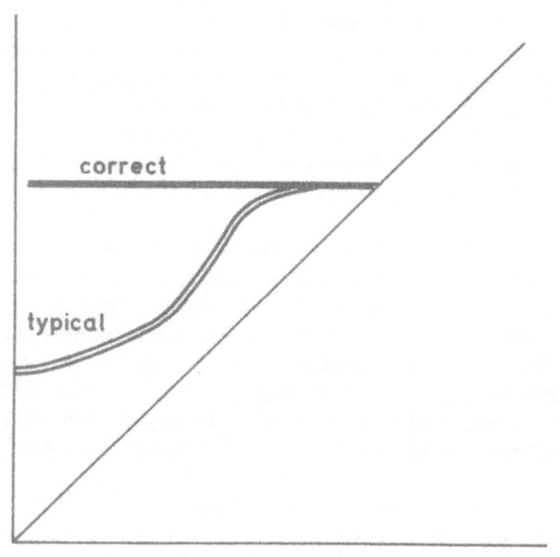

Prediction diagram for finishing a job.
The typical curve is overly optimistic

Predictions of a higher order are also possible. They cannot be represented by a point in figure 5, because more than 2 times are involved. A clever essay by Alan Watts [7], called The Future of Ecstasy, was written in 1974 and pretends to look back from January 6, 1990 to the mid-eighties, let us say to now. Here is a quotation: *"You must understand that despite the ecological crisis of the seventies, technology gave us an enormous amount of leisure. By 1985, there were no longer nine-to-five jobs. The whole world began to run on Greenwich mean time, and work hours today are staggered throughtout each twenty-four hour period, amounting in all about ten hours a week -unless, of course, one is an enthousiast for doctoring, engineering, scientific research, or carpentry, in which case he can work as long as he likes"*. I happen to like science and carpentry and work indeed more than 10 hours a week.

I once tried to make such a reversal myself [8]. NASA had just published a thick book, the NASA Outlooks, 1976-2000 and I tried to imagine how someone, picking this old book from a dusty shelf in 2000, would look at it. Would it sound familiar or funny? Which parts would be obvious and well understood? And which parts would sound strange, historic, dated as a typical mid-seventy stuff? Do not expect answers! The nearest I came was an imagined question of this student to the librarian: "*Here it says that the book deals with space flight for non-military purposes only. What would that have meant? Do you have a historical dictionary telling me what people understood by military at that time?*" You may take this as either a very optimistic or as a very pessimistic comment on my part. Power is power, whether you call it military or not and power in the hands of people with limited perception remains dangerous. I think this comment is relevant to what Lebeau discussed yesterday.

You notice that I have gradually come to look at the long-term movement in history. In a poetic metaphor: we do no longer look at the spray and foam of the sea nor at the short choppy waves but at the hardly noticeable long swell and at the tides. In prosaic mathematical terms: we look at the low-frequency component. Sudden jumps or sudden events easily catch attention and therefore are always the first studied. It is more difficult to notice the presence of a low-frequency wave or warp. This phenomenon is known to carpenters and scientists alike and I have no doubt that it also holds for the study of History. At times it may be very important to get those low-frequency components straight and the extra effort needed may be well spent. This certainly is the experience in doing synthesis radio astronomy, where observations with one large dish or with an interferometer with small dishes may be necessary to get at the small spacings [9].

I immediately knew that I must in this lecture quote O'Shaughnessy's poem The Music Makers [10], which portrays these hardly noticeable but powerful tides in the history of mankind better than anything else I know. It is important enough to project the full text. Please accept it at this moment without comment and for those of you who know, forget the way in which Elgar softened it up when he put music to it. It ought to be tasted raw. I shall come back with some comments in the last section of this talk.

ODE

We are the music-makers,
And we are the dreamers of dreams,
Wandering by lone sea-breakers,
And sitting by desolate streams;
World-losers and world-forsakers,
On whom he pale moon gleams:
Yet we are the movers and shakers
Of the world for ever, it seems.

With wonderful deathless ditties
We built up the world's great cities
And out of a fabulous story
We fashion an empire's glory:
One man with a dream, at pleasure,
Shall go forth and conquer a crown;
And three with a new song's measure
Can trample an empire down.

We, in the ages lying
In the buried past of the earth,
Built Nineveh with our sighing,
And Babel itself with our mirth;
And o'erthrew them with prophesying
To the old of the new world's worth;
For each age is a dream that is dying,
Or one that is coming to birth!

Arthur O'Shaughnessy, 1844-1881

4. Management *

A more prosaic word than management hardly exists. Yet I have to devote a short section to it. A possible attitude in chess is that the only move that ever has to be made is the next move. Similarly, in ESA and NASA the only decisions of consequence are the decisions on the next year's budget. This is the part of the future we are stearing rather than planning. Any longterm planning, including our vision of utopia, serves to lend extra support to make this next move possible in a convincing manner. Practical space policy is deciding on the strategy and tactics for the next 1-5 years, the striped fringe in figure 5. Of course, this includes more than just management but I have taken that one word as a key word. This is a highly technical subject, which is not my field at all. But I have been exposed to it a lot, like anybody who has seen one or more space projects find their miraculous way to completion.

This subject is intimately tied to predicting the future, for signing contracts is an essential part of management. And predicting cost and time of delivery are essential parts of drawing up a contract. We, as scientists or as national delegates in ESA bodies, often feel uneasy about these matters, but we are grateful that others take care of them. I wish to make this gratitude explicit before making some comments that may have a negative tone.

The other day somebody in a TV discussion commented on TV advertising: our housewives are not that stupid. To which the advertising expert answered (abbreviated version): we studied that, yes, they are. I have wondered whether we can change this conversation into one on the promotion of space projects, of which yesterday's movie was a fine example. We might say: Our ministers/delegates are not that stupid. To which the P.R. man might reply: Yes they are. Excuse me for sounding cynical but I understood from Hadot's lecture yesterday that the cynics are respectable philosophers.

Let us look a bit closer at the promotion. I tried at one time to

---

* (this section was omitted from the oral presentation by lack of time).

figure out what keeps the space machine running and came up with the diagram of figure 7, which shows a 4-stage feedback system [11]. Obviously, each of these curves has the tendency to show some saturation, i.e. to curve gently down. But if an equilibrium has been reached, a point at which this machine functions, the curving down means that the equilibrium cannot be shifted upwards. In order to achieve any such shift it is necessary to prove (or to make believe) that at least one of the curves turns decidely upwards just beyond the point where we are now. This has been tried with all four. Just try for yourself and you will soon recognize familiar slogans. You will also recognize the eternal danger that in the absence of sufficient market pull the decisive factor is technology push. For a competent analysis you have to consult other authors (e.g. Harvey Brooks) [12]. Also note that the public response has many layers. Etzioni [13] cites the head line of the front page of the Milwaukee Sentinel of October 5, 1957. "Today we make history". It referred to a game won by the Milwaukee Braves. The news that Sputnik 1 was in orbit was on page 3.

A further comment is that the detailed management system, which is the professed mode of work of a big organization like NASA and probably also ESA, does <u>not</u> really work. The sceptics call it micromanagement. The weakness of the system has received enough publicity since the analysis of the Challenger disaster showed the danger of vital messages from below being suppressed by higer levels. However, plenty of practical experience on less crucial points leads to the same conclusion. Devotion to the work and the wish to see it succeed remain important extra ingredients to make micromanagement successful. It is wise to listen to the ever changing array of fashion words in the NASA management circles. Today, two such fashion words are "tiger team" and "bottom line". Both deny the basic philosophy of micromanagement. The necessity of "tiger teams" denies the assumption that management can be arranged in a perfectly logical, systematic order. "Bottom line" has become a standard way to acknowledge the fact that viewgraphs with many details were once fashionable but now require apologies. The people presenting them say: just look at the conclusion, the bottom line. This attitude may be unavoidable, but clearly undermines the professed theory that every detail has been gone over by independent reviewers. On top of that we have the funny mixture of private companies and civil servants, which is also good for endless amusement or endless quibbling [11].

I must add a final comment about languages, a subject dear to my heart and not one of recent times. For we read in the Bible that the first big project of 'a tower reaching into heaven' (the tower of Babel) flopped on precisely this point. We tend to lament the existence of so may languages in our European community. To cope with this situation takes indeed extra energy and money. But there is also an advantage: far more ofter that the average American we European scientists have experienced situations when perfectly understandable words are not understood, simple because they do not belong to listener's vocabulary. This provides good training for situations with similarly blank faces which regularly occur in joint work among people from different disciplines. The success of many ESA projects (e.g., COS-B, in which people with 5 native languages were involved) shows that such problems can be overcome by care and patience.

Figure n° 7
Promotion of space activities.
The diagram suggests the existence of a 4-stage feedback system.
A raise in the level of effort and funding requires that
at least one of these curves in pushed up

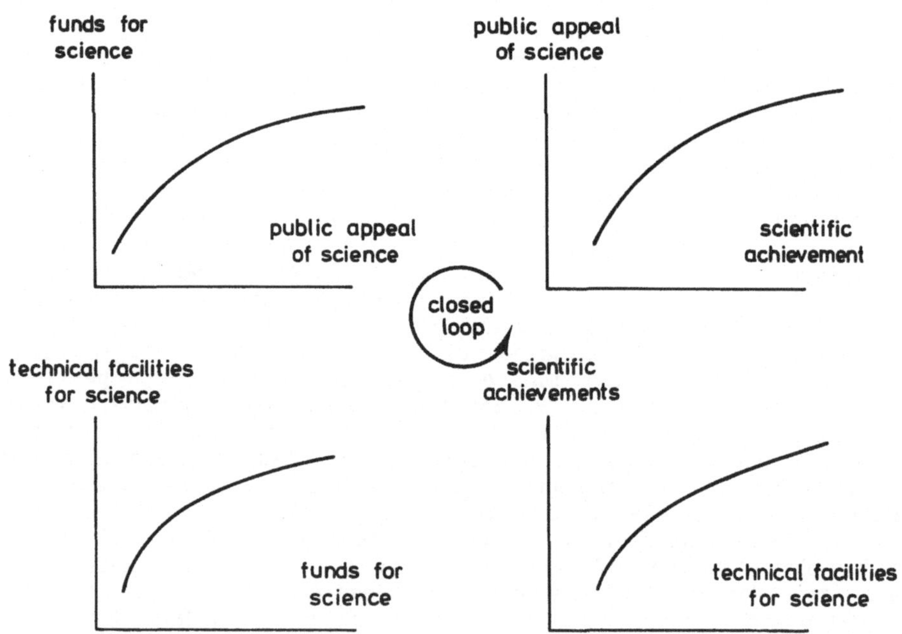

5. Moods

The inspiration for the title of this last section came from a time I was shopping in Princeton. I passed a house in a side street with name plates of about 6 psychotherapists but the surprising thing was the name given to the entire house: Princeton Center for Mood Disorders. This memory came back to me when I was pondering how to address the psychology of man in space. Knowing how vital it is to fully understand and test the various modes in which a space instrument may be working it was tempting to draw the parallel with the various moods in which a person in space may be working. And it is obvious that a full study should include unintended behaviour or malfunctioning, whether you wish to call these failure modes or use some more euphemistic term.

So, to give away my punch line, in my concept of a space station or space town, I do not have the vaguest idea of what there has to be in between but I am sure about the ends. At one end is a Center for Mode Disorders, headed by a man (or possibly a robot). At the other end is a Center for Mood Disorders, headed by a woman. A better concept for inhabitants who insist on Yin-Yang equilibrium I cannot find.

This sets the tone. Now let us look more at ease at a number of aspects, realizing all the time that the only reason why I can speak freely is that I am not an expert and that for anybody in this world the socio-psychology in a space town is a big, big extrapolation from situations he may have studied in detail.

I think we all agree that in big undertakings like space travel we should be cautious and try to build up experience gradually. Therefore, the first legitimate question is: what experience is relevant? I have found references to three kinds.

The first is the experience of explorers or immigrants who ventured into an uncertain future to live elsewhere having to overcome harsh circumstances. Columbus is a much cited example and in the Space Settlement Papers [14] much is made of the parallel with the settlers in Alaska. These parallels do not impress me. Columbus did a lot of technical preparation (some of which was wrong) but he did not have to bring air to breeze. In fact, where he arrived, people were living and he could bring some of them back without great precautions. Settling in Alaska took a lot of know-how, courage, endurance and initiative. Know-how and courage are also needed to settle in space but I guess that endurance would not be of much help and the possibilities for initiative would have to be extremely curtailed in view of all safety precautions.

Can we profit from the experience gained by persons or groups in narrow confinements? Other speakers will address this point. Many studies have been made of those situations, of people trapped in elevators or in submarines. The ultimate study is Sartre's play 'Huis clos' about the people trapped in hell, where even death does not offer an escape. Yet I do not rate the relevance of such analogues very high. The fear to be simply forgotten by the outside world will certainly be less strong. And for technical reasons situations where one person is tempted to save his life at the expense of others are less likely to occur than on a sinking ship.

The third type of experience has been gained by astronauts, payload specialists and passengers on all earlier space flights. This experience must certainly be treasured. I do not know how refined the psychological feedback after those flights has been, among the elaborate health checks. But I am very happy we can hear ourselves during this conference from Hofman [15] and Ockels [16]. However I do have to raise the doubt whether this experience, valuable as it is, would be relevant to space settlers. The question is: are the moods and the possible mood disorders basically different for (a) a person staying on earth, (b) a person spending 5 minutes in parabolic flight, (c) a person going for 5 days in orbit on Spacelab, (d) a person living 5 years or longer in L5, OLGA Town or whatever space settlement may be devised. I think the differences, especially between the last one and the rest, are enormous. The expectancy to return, mentioned by several speakers at this conference, certainly is an important factor. To go into further detail would be mere speculation.

Grounding is a much used concept both in psychology and in psychotherapy. Its opposite "getting high" is even widely used in popular language. To think that getting high can only be achieved by means of drugs has long been recognized as a fallacy. Many years ago I picked up a popular book which had the title (I think, because I lost it) 250 Ways to Get High Without Drugs. I suggest that this is not the right book to give along with an astronaut or space settler. For spinning rapidly and being the center of attention of millions certainly give a high feeling. And while I am here I should also ask about the effect of living through 18 sunrises and sunsets in 24 hours.

So perhaps we should give the space travellers another book along: 250 ways to ground (again without drugs). I already know two ways which seem impractical in these circumstances: walking with bare feet through the wet grass and lying down quietly enjoying your one weight. This all may sound rather lighthearted but I am dead serious. I even have wondered how funerals will be arranged in space, at least in those circumstances in which the survivors have the technical possibility of a choice. Perhaps that point will be taken up in a discussion.

On a more happy note, I had the privilege and pleasure to chair the first meeting where an American and a Russian space travellers got together. One was called an astronaut, Glenn and the other a kosmonaut, Titov. The meeting was on earth and more exactly at Washington, in 1962. I was COSPAR president at the time. Two points about this event are worth mentioning in the present context. First, a strong consideration in agreeing that this meeting should be held under COSPAR auspices, where it hardly belonged , was that there had not yet been a major disaster. It was politically important to have such a meeting in a relaxed atmosphere and a disaster at one of the sides would have made this impossible for a long time to come. Secondly, I presented a very short speech at the end, given them each a wooden shoe form a pair I had bought in my home town. Grounding must already then have been on my mind, although I had not yet heard of the word, except in electricity.

I now like to go back to the Music Makers. Have you noticed the line "*sitting by desolate streams?*" It has a very double character. For although it contains the word desolate, it is quite reassuring. Both "*sitting*" and "*streams*" have such a grounding quality. Dreaming at a stream is healthier than just dreaming of a stream and never being in touch at all with a stream would be terrible. This would be only one of the hardships to the moods of space inhabitants.

Clearly, we have only begun to make guesses. Perhaps much wiser persons than we can help us discover the way. Perhaps thinking that we needed a drink to feel good, need to knock on wood for good luck to continue, and need a touch of earth to feel grounded are all equally primitive. The buddhist teacher Tarthang Tulku [17] shows that grounding in space is quite possible.

Notes and references

[1] Some material in this section and the next is drawn from a lecture I presented in a symposium on The Future of Science in Space at Philadelphia, June 1976, which -to my knowledge- has not been published.

[2] For instance: J. Lighthill, 1986, The Recently Recognized Failure of Predictability in Newtonian Dynamics, Proc. Roy. Soc. A407, 35-50

[3] First presented in H.C. Van de Hulst, 1983, "Comments", in The First 25 Years in Space, A.A. Needell, ed., 107-114, Washington D.C., Smithsonian Institution Press

[4] H. Friedman, 1965, The Next 20 Years of Space Science, Astronautics and Aeronautics, 3, 40-47

[5] P.R. Escobal, Methods of Orbit Determination, 1965, reviewed by H.C. Van de Hulst in Space Science Reviews, 6, 414, 1966

[6] H. Van de Wall, 1975, Falsifications in Art

[7] Alan Watts, 1974, Cloud-Hidden, Whereabouts Unknown, New York (Vintage Books)

[8] Taken from reference 1.

[9] For a fuller explanation of the metaphor, see H.C. Van de Hulst, 1984, Nanohertz Astronomi, in The Early Years of Radio Astronomy, W.T.Sullivan, ed., 385-398, Cambridge, Cambridge University Press

[10] Arthur O'Shaughnessy, Ode, cited from the Albatross Book of Golden Verse, 1933, 478-479.
I am still grateful to the teacher who put this recently appeared anthology on the mandatory book list.

[11] Taken from reference 1

[12] H. Brooks, 1983, The Motivation for Space Activity, in The First 25 Years in Space, A.A.Needell, ed., 6-26, Washington D.C., Smithsonian Institution Press and literature cited here

[13] A. Etzioni, 1983, Comments in The First 25 Years in Space, A.A. Needell, ed., 33-36, Washington D.C., Smithsonian Institute Press
[14] Eric Jones, editor, 1986 The Space Settlement Papers, J. British Interplanetary Interplanetary Society 39, 29-311
[15] J. Hoffman, 1987, in this volume
[16] W. Ockels, 1987, in this volume
[17] Tarthang Tulku, 1977, Time, Space, Knowledge, a New Vision of Reality, Emeryville, California (Dharma Publishing)

Interventions

Rudolf Albrecht

A comment more than a question: in the first part of your lecture, when you gave examples of extrapolations, you made the statement that extrapolation is possible in "mathematically well defined models". This contains the assertion that it is possible to develop models, which approximate reality. The qualifier is that the models are mathematical, i.e. that Mathematics can be used as a tool to manipulate the model. This implies, however, that results can only be derived within the constraints of the tool, of mathematics. Mathematics itself is an internally consistent system of logic, resting on certain axioms. For example, many proofs of mathematical theorems are done by disproving the contrary, discussing the validity of the "Tertium non datur". However, one can build systems based on other axioms, thereby changing the tool of manipulation. This has to be kept in mind when making extrapolation using mathematical models.

Eric M. Jones

In the space community, predictions of events can be accurate about form, but inaccurate about time of occurrence.

LOIN DE LA TERRE :
UNE NOUVELLE PHASE DU "REFOULEMENT ORGANIQUE" ?

Gérard Huber
Psychanalyste
8, rue Geoffroy Saint Hilaire 75005 Paris, France

Summary

FAR FROM THE EARTH:
A NEW STAGE IN THE "UNCONSCIOUS ORGANIC REPRESSION?"

Man's conquest and inhabitation of the space encounter a lot of technological aspects, technological solutions will not fail to answer. However, the latter will induce drastic situations which will lead philosophy and the human sciences to throw light on the stakes as regards the individual, society and human race. In such an experimental macro-laboraty that represents the space adventure, the hard and soft sciences have to find a common language allowing to deal together with the future of mankind.

The narcissist omnipresence of technology generates an homogeneous subject who maintains to act in the interest of the individual, society and humanity. Enthusiasm is supported by the innermost conviction that it is lead by the idea of the principle of the "Good" of humankind.

However, a splitting of the thechnological monolithism is to be observed, Tchernobyl being not the less representative name, in the nuclear field, neither Challenger in the spatial one. It is the reason why NASA, unwilling to limit its self-affirmative pulsions, organizing as a supershow the shuttle departure although many negligences, among which some were well known, have been reported since, felt through a faulty act.

In fact, technology is so abhuman that man is compelled to risk catastropher just to survive.

As far as spatial conquest is concerned, the shock will be characterized by such a privation that it will constrain man to drastic changes in his way of being and in his organization.

In other words, mankind neglected to think about his evolution: it is only now, at the time technique ought to ensure the future of the mankind and be only a tool for his well-being, that this question is faced.

"Homo technicus" and the soon to come "Spatiopithec" [1] are on the verge to lose Earth and Man without getting ready to think Earth and Man as lost objects.

J. Schneider and M. Léger-Orine (eds.), Frontiers and Space Conquest, 87–97.
© 1988 by Kluwer Academic Publishers.

Psychoanalysis is well familiar with the idea of lost object -whether it be a physically elaborated reality by identification or a psychological one independent of any relationship to any real representation. Psychoanalysis may help us conceiving "Tele-Earth" and "Tele-man". Furthermore, it can also put inlight on the way mankind will develop in an environment characterized by microgravity and the loss of the verticality.

Indeed, according to Freud, it is thanks to his verticalization that man owes his ability to elaborate his way of civilization. This evolution is observed together with an "unconscious organic repression" shifting from an erotic olfactory anal stage characteristic of the Homo period to an erotic visual phallic one.

This raises the following question: will the exploration and inhabitation of the space lead not the erotic visual phallic stage to die at this determinant period of the "unconscious organic repression"?

The consequences would then be more drastic. The erotic visual phallic stage being impregnated of the development of sexuality, its disappearance might be in the end of the sexuality itself and, then, by the dying out of the human species itself.

Psychoanalysis concentrates on the following question: let us imagine that "spatiopitecs" can survive and reproduce in space: will their lives not be devoid of authenticity? Are the necessary steps taken in order to prevent them to become "uninhabited inhabitants" of space?

La conquête et l'habitation de l'espace se heurtent à maints problèmes techniques. Les réponses techniques ne manqueront pas. Mais elles confrontent l'humanité -définie comme sujet qui a en charge le destin de l'espèce humaine- au devenir de l'hominisation même.

Ce qui se vit d'ores-et-déjà dans les laboratoires expérimentaux comme un nouvel exode, à savoir la libération du champ gravitationnel, s'annonce comme la voie royale qui mène à une modification essentielle du comportement humain, voire du genre humain et, pourquoi pas, de l'espèce. Dans ce macrolaboratoire sciences de la matière inorganique, sciences du vivant non-humain et sciences du vivant humain viennent non seulement s'<u>inter</u>-discipliner, mais s'<u>intra</u>-discipliner.

L'humain, comme objet scientifique à venir, devient un objet de transmission, au sens fort du mot.

Le paradigme scientifique me paraît devoir être formulé en ces termes : Comment transmettre l'humain dans un champ caractérisé par la microgravité, puis dans un champ gravitationnel reconstitué ? A ce paradigme correspond le paradigme éthique, que je formulerai en ces termes : Faut-il transmettre l'humain ? Autrement dit, quel droit l'humain a-t-il à être transmis ?

C'est, en effet, au moment où l'humanité se rend compte qu'elle ne dispose pas d'un savoir sur la transmission de l'humain susceptible d'être appliqué et de déboucher sur un apprentissage, que se pose, pour elle, la question de cette transmission. C'est là, dans ce type de questionnement, que se rencontrent la pensée épistémologique et la pensée éthique, là, dans un dialogue préalable et originel, même si celui-ci peut apparaître comme constitué après-coup. Ce dialogue contient la question de l'espèce d'espace que nous voulons, celle de l'espèce que l'espace requiert, celle de l'espèce que nous voulons -contre l'espace- pour l'espace.

Ce double questionnement ne manque pas de se faire sur fond de "tératomorphie", au sens de "*la modalité selon laquelle, dans l'évolution de la vie animale, apparaissaient les mutations aboutissant à la constitution de nouvelles espèces jusqu'à l'homme y compris*" [2].

- Sur le plan phylogénétique, la tératomorphie est l'idée que les formes nouvelles qui correspondent à la sélection naturelle et à l'adaptation des espèces, autrement dit aux pulsions d'auto-conservation, sont perçues comme des matérialisations de la mort (voire même du meurtre) par les espèces qui ne peuvent changer, et qui succombent, quant à elles, aux pulsions de mort, soit en disparaissant (et là il y a victoire du meurtre), soit en se fixant.

- Sur le plan ontogénétique, la tératomorphie est le format que reçoivent les perceptions de pulsions à l'intérieur même du processus de transmission d'une génération à l'autre.

Ainsi, lorsque l'espace est perçu comme un gisement à exploiter dans sa triple dimension d'apport énergétique, de réservoir de matières premières et d'espace physique [3], il est mis en position de Cause, c'est-à-dire de Source de pulsions d'auto-conservation de l'espèce humaine (et des espèces domestiquées). Mais ces pulsions auto-affirmatives sont susceptibles d'être vécues, comme c'est le cas dans la transmission d'une génération à l'autre, comme des pulsions de mort par la génération qui suit. La catastrophe dans laquelle a été prise la destruction de Challenger nous servira ici d'exemple. En organisant à grands coups de spectacle médiatique le départ de la navette, malgré les négligences nombreuses qui ont été répertoriées depuis, la NASA a commis un acte manqué qui possibilise l'acceptation ultérieure d'une intrication étroite et non maîtrisée des pulsions d'autoconservation et des pulsions de mort par la population de la Terre.

Le traumatisme qui s'en est suivi dans la jeunesse tout particulièrement est un indicateur de ce que peut représenter l'affirmation -même réussie- des pulsions de vie d'une génération par celle qui suit.

Ce que Freud a appelé la "toute-puissance infantile", c'est précisément cette essence auto-affirmative et illimitée des pulsions de vie qui consiste à ne pas laisser la moindre place à celles des autres, c'est cet effort dément d'être un sujet homogène. Nous n'avons pas ici à développer les connexions de signification que nous pouvons trouver, lorsque nous comparons la conquête de l'espace à la construction de la Tour de Babel. Qu'il nous suffise de dire que la Babel moderne est surdéterminée par l'illusion de la reconstruction d'une origine perdue, c'est-à-dire par la dénégation langagière de cette perte dont le langage a fait son propre objet.

En rapprochant la division des langues telle qu'elle s'est pratiquée dans le récit biblique de l'éclatement du monolitisme techno-logique qui se produit çà et là sous nos yeux (je pense naturellement à Challenger, mais aussi à Tchernobyl), je veux souligner que, dans la transmission du sujet de la recherche et de la science, c'est en amont et non en aval que doivent se penser les conditions d'un retrait des pulsions auto-affirmatives susceptible de laisser être celles des autres (que ceux-ci soient d'autres chercheurs, le prétendu public, ou les générations à venir).

Ce retrait est en même temps une pratique du partage et implique une acceptation de la structure dissymétrique qui met l'autre en ma présence. Si le chercheur pense, c'est à bien des égards, et ce, malgré ses discours sur la différence et la variété, dans une perspective normée par la substitution du sujet à l'autre, ce qui le rend très proche d'un mode de pensée opératoire.

En développant la phrase que j'ai citée plus haut dans un texte sur La folie du Jour de Maurice Blanchot, Emmanuel Lévinas avait-il à l'idée que la vérité de ce constat commençait avec l'homme, c'est-à-dire au moment même où l'homme se soucie d'hominiser l'hominisation ?

Tel est peut-être le véritable vertige de la technique.

La technique est à ce point abhumaine qu'elle commande à l'homme de courir à la catastrophe, au sens où l'entendait Sandor Ferenczi d'un "choc exogène", pour survivre.

Ce choc se caractérisera par une privation telle que l'homme sera contraint de changer ses modes de fonctionnement et son organisation.

Se référant à la période de l'assèchement des océans, Ferenczi fait cette hypothèe : "*Nous avons soutenu que ces êtres se sont certes adaptés à la nouvelle situation, mais avec l'intention secrète de rétablir l'ancienne situation de quiètude dans ce nouveau milieu le plus vite et le plus souvent possible*" [4].

Ce questionnement, partagé par Freud, est normé par une théorie de l'évolution. Mais cette théorie n'est pas darwinienne ni lamarckienne. Avec Freud et Ferenczi, "*il ne suffit pas de remarquer que l'espèce est en lutte avec le monde extérieur, d'évaluer les triomphes sélectifs de l'organisme, mais il faut encore resituer le développement libidinal, en partant du fait que l'individu est en conflit avec le monde intérieur, et tenir compte des triomphes sélectifs du refoulement*" [5].

Je reviendrai sur la question du devenir de la libido et des pulsions. Pour l'heure, je soulignerai en quoi ce questionnement permet de lire ce qui se prépare, non plus dans le secret, mais à ciel ouvert, avec la fondation d'une civilisation spatiale.

Sachant qu'il existe grosso modo quatre degrés de pénétration de la biosphère -l'exploration, l'occupation à des fins d'étude, l'occupation à des fins d'exploitation et enfin la colonisation- nous comprenons immédiatement que la maîtrise de la microgravité porte en elle la possibilité d'une extension de l'anthroposphère au-delà de la biosphère [6].

Cette pénétration a pour condition le renoncement progressif à l'influence des forces gravitationnelles et, au-delà, le renoncement à la victoire de l'espèce humaine sur la pesanteur.

C'est dans ce contexte d'ambivalence vis-à-vis de la microgravité que la technique doit produire le corpus de ses réponses spécifiques qui accompagnera l'homme et le modifiera progressivement à tous les niveaux de son organisation, dès la construction de la plus petite cellule technique comme, par exemple, celle du Laboratoire expérimental sur terre, ou celle de la station occupée à des fins d'étude (Saliout, Skylab). Autrement dit, à quelque degré que ce soit, la technique a pour but immédiat (au sens où la dichotomie traditionnelle entre la fin et les moyens est périmée dans l'idéologie de la recherche spatiale, dans l'idéologie de la technique, même si elle garde une certaine signification dans la programmation hic et nunc) de transférer toutes les réponses techniques de la Terre sur le nouveau territoire spatial construit ex nihilo. (On notera la nouvelle absurdité du concept de territoire, qui devient, de la sorte, un horizon qui subira en son temps, une modification sémantique). C'est dans ce transfert que la technique se trouve en rupture avec elle-même. Tout d'un coup se pose, pour elle, la question d'être hominisatrice, au sens d'une hominisation de l'hominisation, et c'est à ce moment que les tenants de la technique se rendent compte qu'elle a oublié la question de l'humain en route.

Tout d'un coup elle doit prendre le relais des organisations complexes qui biophysiologiquement ont initié l'hominisation en instaurant la verticalité pour la nouvelle espèce. Dans l'espace, le bio-physio-psychologique ne peut rien. Il doit déléguer son pouvoir à la technique.

Dans la perspective encore très éloignée de la colonisation de l'espace, de la construction de villes habitables par notre espèce, il y va de la possibilité pour la technique de rétablir, après l'avoir perdu, le champ gravitationnel, de restaurer la verticalité. Mais rien n'indique que ce rétablissement soit programmé. C'est là que se situe la question de la création d'une nouvelle espèce.

Quoi qu'il en sera, il faudra commencer par penser l'objet -ici la verticalité- comme ayant été perdu. Sinon c'est la trace de l'effacement de la trace de la verticalité qui sera perdue. Mais peut-être sera-ce la condition sine qua non de l'émergence d'une nouvelle espèce.

Dans la situation d'explorateurs de la biosphère et dans celle de la sortie de la biosphère, c'est donc la Terre qu'il convient de penser d'ores-et-déjà comme objet perdu. La technique nous oblige à la penser comme une "Téléterre".

Certes la "Téléterre" n'est pas un objet actualisable immédiatement (sauf pour quelques cosmonautes qui incarnent un autre visage de l'humain). Mais ceci n'est pas l'essentiel, lorsque l'on constate que la pensée techniciste n'est plus normée par le couple actuel/possible. Contrairement à ce qui prévalait dans la métaphysique aristotélicienne qui accompagnait encore la science européenne jusqu'au XIXème siècle, l'actuel est plus proche du possible que de l'être.

Au passage notons que c'est sans doute pour cela qu'Heidegger a dénoncé la souveraineté de la technique, en quoi il voyait une menace pour l'être auquel se substitue le possible, c'est-à-dire l'annonce d'un oubli définitif de l'être et de la question qu'il porte.

On sait que la critique de Lévinas porte précisément sur l'ontologie première de Heidegger, ontologie qui, se recentrant sur l'être, oublie le Même et la question qu'il porte. Autrement dit peu importe de dénoncer qu'il n'est plus possible de penser l'être, si cela signifie être prêt à payer de l'impérialisme du Même le retour de l'être. Et Lévinas d'affirmer que l'opposition de Heidegger à la passion technique, "*issue de l'oubli de l'être caché par l'étant, demeure dans l'obédience de l'anonyme et mène, fatalement, à une autre puissance, à la domination impérialiste, à la tyrannie*" [7].

Appliqué à la question qui nous occupe, le questionnement lévinassien m'anène à considérer que l'actualité de la "Téléterre" équivaut à l'actualité de son devenir-autre. Autrement dit il n'y aura (retour à l'être) de "Téléterre" qu'à la condition que la Terre soit autre. Traduit dans un langage socio-idéologique, cela signifie qu'une des conditions de l'habitation de l'espace est que les conditions d'existence terrestre aient été au préalable modifiées dans le sens d'une intégration de l'altérité.

Il y a donc actualité de la Terre/objet perdu, et cette actualité tient à ce que cette perte est le possible même de la conquête de l'espace, le possible au sens de processus cognitif auto-organisateur technico-technique. La connaissance de l'<u>espace est processualisation de la perte de la Terre</u>.

C'est cette processualisation qui donnera lieu aux conflits entre générations. La Grande Cause, celle qui sera la raison même de la conquête spatiale, se constituera comme Vérité abominable, parce que l'habitation de l'espace sera vécue comme l'expression d'une infidélité autrement humaine que l'infidélité dont nous parle Hölderlin, lorsqu'il l'oppose à l'infidélité divine [8].

L'infidélité comme trahison incestocratique. La Terre comme horizon mondial des possibilités n'aura succédé qu'un court moment à Dieu comme sujet permettant à l'homme de se raconter lui-même.

Une question se pose : au cours de cette processualisation, les éléments organiques vont-ils manquer l'occasion qui leur est offerte de mourir ?

Je voudrais, pour penser cette question, me référer à l'hypothèse freudienne d'un "refoulement organique".

Le noeud de cette hypothèse est qu'à un certain stade du processus d'homonisation, notamment au stade déterminant du début constitutif de ce processus, l'actualisation des pulsions antérieurement dominantes et devenues secondaires du fait du franchissement opéré par l'organisation qui se développe exige un tel effort qu'elle ne peut pas ne pas s'accompagner de représentations marquées de déplaisir, que le psychisme doit, pour cette raison refouler.

L'organique n'est pas seulement, comme chez Hall [9] le niveau infraculturel sur lequel la dimension culturelle a un effet de feed-back, c'est le niveau biopsychique de la réalité humaine appréhendée dans ses dimensions phylo- et ontogénétique. Le "refoulement organique", c'est l'idée que les pulsions du corps, envisagé comme une organisation complexe somatico-érotique, subissent un refoulement primaire aux étapes déterminantes du processus d'hominisation.

Freud donne un exemple précis de refoulement organique dans Malaise dans la Civilisation. S'interrogeant sur la fondation de la civilisation humaine, et plus particulièrement, sur celle de la famille, Freud s'attarde sur la périodicité du processus sexuel et sur son influence quant à l'excitation sexuelle psychique. Il écrit : "*En réalité la périodicité du processus sexuel s'est maintenue, mais son influence sur l'excitation sexuelle psychique a tourné plutôt en sens contraire. Ce revirement se rattache avant tout à l'effacement du sens de l'odorat dont l'entremise mettait la menstruation en état d'agir sur l'esprit du mâle. Le rôle des sensations olfactives fut alors repris par les excitations visuelles. Celles-ci contrairement à celles-là (les excitations olfactives étant intermittentes) furent à même d'exercer une action permanente. Le tabou de la menstruation résulte de ce "refoulement organique" en tant que mesure contre le retour à une phase surmontée du développement*". Puis "*cependant le retrait à l'arrière-plan du pouvoir excitant de l'odeur semble être lui-même consécutif au fait que l'homme s'est relevé du sol, s'est résolu à marcher debout, station, qui rendant visible les organes génitaux jusqu'ici masqués, faisait qu'ils demandaient à être protégés, et engendrait ainsi la pudeur. Par conséquent le redressement ou la "verticalisation" de l'homme serait le commencement du processus inéluctable de la civilisation*". Remarquant que l'impulsion à être propre procède du besoin impérieux de faire disparaître les excréments devenus désagréables à l'odorat, Freud précise que "*l'érotique anale succombe la première à ce "refoulement organique" qui ouvrit la voie à la civilisation*" [10].

Ainsi la victoire sur la pesanteur s'accompagne d'un déplacement de la centralité pulsionnelle et érotique des perceptions oflactives auxquelles est liée l'érotique anale aux perceptions visuelles auxquelles est liée l'érotique phalllique.

Mais Freud va plus loin. Il ajoute : "*Du fait du redressement vertical de l'être humain et de la dévalorisation du sens de l'odorat, non seulement l'érotique anale mais bien la sexualité tout entière aurait été menacée de succomber au refoulement organique...Il s'attesterait ainsi que la racine la plus profonde du refoulement sexuel, dont les progrès vont de pair avec ceux de la civilisation, résidât dans les mécanismes organiques de défense auxquels la nature eut recours, au stade de la station et de la marche debout, en vue de protéger le mode de vie établi par cette nouvelle position contre un retour du mode précédent d'existence animale*" [11]. Dans ce scénario, l'érotique phallique a hérité du devenir de la sexualité -et partant de la reproduction- à laquelle la nouvelle espèce était tout près de renoncer. C'est pourquoi, dans le dispositif analytique, la visualisation des mots est une issue au refoulement sexuel et au refoulement de la sexualité.

La question qui se pose, avec l'exploration puis l'habitation de l'espace, est de savoir si l'érotique phallique/scopique ne va pas succomber à son tour à ce nouveau stade déterminant du "refoulement organique" caractérisé par l'effacement de la pesanteur et des traces mnésiques de la victoire de l'homme sur la pesanteur. A nouveau se présenterait l'éventualité d'une disparition de la fonction sexuelle, doublée, bien sûr, de celle d'une disparition de l'espèce humaine.

Ce questionnement correspond à l'hypothèse que Freud formule en ces termes dans Pourquoi la Guerre ? "*Je pense la chose suivante : depuis des temps immémoriaux, le processus de l'évolution de la culture se déploie sur l'humanité (Je sais que d'autres préfèrent l'appeler : civilisation). C'est à ce processus que nous devons le meilleur de ce que nous somme devenus et une bonne partie des maux dont nous souffrons. Ses motifs et ses origines sont obscurs, son issue incertaine, certains de ses caractères aisément repérables. Peut-être mène-t-il à l'extinction de la race humaine, car il porte atteinte à la fonction sexuelle à plus d'un égard, et dès aujourd'hui les races incultes et les couches arriérées de la population s'accroissent plus fortement que celles qui sont très cultivées. Peut-être ce processus est-il comparable à la domestication de certaines espèces animales ; il entraîne sans aucun doute des modifications corporelles ; on ne s'est pas encore familiarisé avec la représentation selon laquelle le développement culturel est bien un tel processus organique*" [12].

La civilisation spatiale sera ou ne sera pas. Quel que soit son degré de haute technicité, elle se révèlera dans sa réussite comme dans son échec, un stade spécifique -accompli ou empêché- du processus organique que constitue l'auto-organisation individuelle et collective des humains. Réussi, le refoulement -sur lequel nos hypothèses ne sauraient être que limitées- permettra à cette civilisation d'accéder à une fondation. Impossible -au-delà de tout deuil d'ailleurs-, ce refoulement demeurera l'objet qui se dérobe devant toutes les générations qui seront confrontées à cette fondation.

Une autre question se pose : le système techno-logique sera-t-il apte à traiter l'expérience vécue par les fondateurs ? Cette question concerne d'ores-et-déjà le "microtechnocosme" que constituent et constitueront les vaisseaux spatiaux et les stations orbitales.

Ce "microtechnocosme" est une microsociété dans laquelle le sujet n'appréhende pas l'autre directement et de manière concrète, mais artificielle, c'est-à-dire par l'entremise obligée d'artefacts conceptuels et de robots matériels. Dans ce milieu, dont toute authencité, au sens de Lévi-Strauss, semble exclue, "*milieu technique total extrêmement sophistiqué et radicalement coupé du milieu terrestre originel (la seule relation existante demeurant -outre le contact radio tout à fait marginal- la communication entre les ordinateurs de bord et les ordinateurs de base sur Terre)*" [13], l'homme devient un animal domestique, qui transfère sa souveraineté à la technique et se voue à des rapports opératoires-techniques exclusifs.

Dans ce contexte, il est légitime d'attendre un remaniement pulsionnel concomitant à une réduction de la complexité psycho-sexuelle de l'individu.

La Cité technique semble se réserver aux individus capables de performances hypermnésiques, c'est-à-dire à ceux qui sont capables de fonctionner à l'abri des déformations qu'impliquent la sexualité psychique et l'infantile, en harmonie avec une situation d'isolation. On peut imaginer que la stabilité de la coupure dans le tissu associatif psychique qui sera requise chez ces futurs spatiopithèques sera telle qu'elle mènera rapidement à l'effacement de l'opposition de deux modes de pensées : "*la pensée personnalisée, branchée sur l'Inconscient et inscrite dans le temps et l'histoire du sujet, pensée illogique traversée par le fantasme, et la pensée impersonnelle, coupée de l'Inconscient, pensée logique et performante à l'abri des fantasmes, "pensée opératoire"* " [14]. Mais la psychanalyse nous apprend que là où le système logique est inapte à traiter l'expérience vécue, il y a émergence d'une névrose d'angoisse et d'une angoisse automatique. On sait que la technique, de nature chimique, électrique ou électrochimique, s'efforce de prévoir cette situation. La découverte des neurotransmetteurs, molécules sélectives de transmission intracérébrale, permet d'envisager une manipulation de "*l'expérience interne*" [15] telle que les promesses s'avèrent nombreuses.

Cependant une question demeure : quelle sera la mesure de l'aptitude du système technologique à traiter l'information liée à l'expérience interne ?

La psychanalyse s'est donnée une mesure : le refoulement. C'est d'ailleurs pourquoi la levée du refoulement -qui n'est jamais définitive- permet au sujet d'accéder à la complexification de sa vie propre et au traitement de l'information qui lui est liée.

Mais qu'en est-il de la techno-logie ? En apportant une autre réponse à la question de l'effet en retour de la culture sur la fonction sexuelle, celle d'une nouvelle domesticité de l'homme, que produit-elle qui puisse valoir comme mesure de cette expérience ? Il est possible de poser la question autrement : En quoi les futurs spatiopithèques, selon le néologisme introduit par Jean Schneider, seront-ils encore habités par l'expérience de leur vécu et conscients de l'être ? Nous nous souvenons ici de l'invitation au voyage de Léonard de Vinci : "*Toi qui veux voir comment l'âme habite le corps, tu n'as qu'à regarder comment le corps use de sa quotidienne habitation*".

Mais le spatiopithèque aura-t-il encore la sensation de son corps ? Son corps ne sera-t-il pas dissous dans l'espace, devenu le vrai lieu d'habitation de l'artefact ?

Références

[1] Le Spatiopithèque : vers une mutation de l'Homme dans l'Espace, A. Brahic et al., Le Mail, Editions Payot, 1987
[2] Roger Laporte et Bernard Noël : Deux lectures de Maurice Blanchot (Fata Morgana, Montpellier). Cité par Emmanuel Lévinas in Sur Maurice Blanchot, Fata Morgana, Montpellier, 1975, p. 57
[3] Lire à ce sujet L'espace en héritage, d'André Lebeau, Paris, Editions du Seuil, 1986, p. 48-49
[4] in Thalassa, Essai sur la théorie de la génitalité, in Psychanalyse 3, Paris, Payot, 1982, p. 318
[5] Patrick Lacoste, Destins de la Transmission, in S. Freud, Vue d'ensemble des névroses de transfert, Paris, Editions Gallimard, 1985, p. 179
[6] A. Lebeau, in op. cit., p. 289.
[7] in Totalité et Infini, Martinus Nijhoff, La Haye, 1974, p. 17
[8] On lira avec profit la préface que Jean Bauffret a écrite sur ce sujet in Holderlin/ Remarques sur Oedipe, Remarques sur Antigone, Bibliothèque 10-18, Paris 1965.
[9] in La dimension cachée, Paris, Seuil, 1971
[10] Paris, PUF, 1971, p. 49-50
[11] Ibid., p. 58
[12] in Résultats, idées problèmes II., PUF., 1985, p. 214-215
[13] Gilbert Hottois, in Le signe et la technique, Paris, Aubier, 1984, p. 98
[14] Christophe Dejours, in Le corps entre biologie et psychanalyse, Paris, Payot, 1986, p. 7
[15] Gilbert Hottois, Ibid. p. 97

Interventions

Vana Secardin

Heidegger annonce la technique comme un oubli définitif de l'être. Ceci devrait être la base d'une considération importante concernant la conquête spatiale. L'home sera totalement dépendant de la technique, des instruments. S'ils lui manquent, il est perdu car la survie de l'être humain est impossible dans l'espace, milieu totalement inhospitalier sans intermédiaire pour lui.

Une autre réflexion de type métaphysique pourrait porter sur la considération de la conquête spatiale comme un acte démiurgique de la race humaine pour échapper à son destin de destruction actuel.

Jean Heidmann

Vous avez souvent évoqué la perte de la Terre ou de la verticalité. A court terme, une génération par exemple, ceci est peut-être important. Mais à plus long terme il n'en sera pas toujours ainsi : la Lune et Mars, avec leurs gravités, seront occupés. De plus, même en vol orbital, on pourra créer des gravités artificielles si cela est avantageux. Et aussi revenir sur Terre. Il faut donc prévoir une adaptation de l'homme à la non-pesanteur *et* à la pesanteur. Peut-être, plutôt que d'*Homo ingravitus*, devrait-on parler d'*Homo agrativus* pour le prochain siècle.

LA CONDITION PRESPATIALE
MANIFESTE POUR UNE ANTHROPOLOGIE SPATIALE

Jean Schneider
Observatoire, 92195 Meudon, France

Nous sommes donc ici pour réfléchir à divers aspects philosophiques des techniques spatiales, au spatial, comme on dit plus brièvement ou, par un glissement supplémentaire, à l'Espace. Ainsi est-ce une occasion d'amorcer par ce biais particulier de la technique spatiale un dialogue entre les sciences et la philosophie. Il y a certainement d'autres points d'entrée que le spatial pour un tel dialogue, mais enfin c'en est un. Ce dialogue prend généralement la forme de l'épistémologie ou si l'on est plus exigeant, de la philosophie critique qui exerce son talent d'analyse sur la démarche de la science. C'est ainsi qu'à propos du spatial on peut relever que dans la démarche qui permet de voir la terre d'en haut la raison précède l'oeil réel.

Cette captation des activités spatiales par la philosophie critique est d'ailleurs le germe d'un premier débat intéressant, voire d'une contradiction qu'il ne faudrait pas étouffer et que l'on peut tenter de schématiser ainsi : la philosophie vise à élaborer un discours sur les activités spatiales à partir d'une position critique dont l'ancrage terrien est déclaré être accidentel ; d'un autre côté, c'est clair dans l'idée de "cosmicisation", agrandissement de celle "d'hominisation" développée par A. Ursul, c'est le mouvement inverse qui a lieu : c'est à partir de l'Espace que s'élaborera un discours sur la philosophie critique d'aujourd'hui. Les philosophes critiques ont beau jeu de remarquer qu'une telle déclaration est, de fait, proférée à partir d'une base purement terrienne-logocentrée ; cette argumentation est aussi inattaquable que celle de Zénon qui démontre que le mouvement ne peut pas exister.

Il me semble que dans une telle situation nous devrions accorder à ces impossibles discours prospectifs, discours sur l'après discours, une attention flottante ouverte.

Ainsi le spatial, ou l'Espace, n'est pas seulement une technique. Par les quelques remarques que je vous propose je voudrais développer cette idée.

Comme première tonalité non technique de l'espace, je mettrais en relief ce qui a déjà été évoqué de-ci de-là dans ce Colloque, principalement par Maria Villela-Petit, je veux dire notre espace intérieur, celui où peut se déployer l'imaginaire, que ce soit par la technique de la narration, ou par l'art par exemple, et qui fait notre richesse spécifique d'humains.

J. Schneider and M. Léger-Orine (eds.), Frontiers and Space Conquest, 99–103.
© 1988 by Kluwer Academic Publishers.

La technique des propulseurs nous permet d'envoyer dans l'espace la fine fleur de la technologie, que ce soient la sonde automatique Viking sur Mars, les sondes Giotto et Vega vers la comète de Halley, ou la sonde Voyager vers Uranus et Neptune -sonde dont vous savez qu'elle a pu être reprogrammée en cours de route par commande à distance. Mais veillons à ce que ces puissantes fusées n'oublient pas d'envoyer aussi dans l'espace la fine fleur de notre espace intérieur, ou si vous voulez de l'esprit.

Voilà que par cette recommandation morale nous passons du domaine de l'épistémologie et de la philosophie critique à celui de l'éthique. Celle-ci nous commande, dans le contexte spatial, une sorte de devoir cosmique qui serait celui-ci : allons loin dans l'espace, oui, le plus loin possible, mais pas n'importe comment, pas au détriment de la qualité spirituelle.

Mais l'esprit, notre espace intérieur qui au fond devrait avoir ses propres fusées internes, dépend toujours d'un corps. Et notre corps, aussi bien H. Oser que M. Villela-Petit nous l'ont rappelé, est tributaire des conditions terrestres qui l'ont façonné. Ecartons-nous de ces conditions et notre corps sera autre. Et donc aussi, peut être, notre psyché si nous admettons cet axiome d'un lien de son interdépendance au corps. En matière de lien psychosomatique on ne peut s'adresser à meilleure enseigne qu'en se tournant vers la psychanalyse.

Plusieurs pistes sont suggérées par cette discipline : il y a d'abord le rapport à la verticalité dont la connotation parternelle est multiple, qu'il s'agisse du terme "érection" ou d'Ouranos qui est au-dessus de nos têtes. L'intérêt pour l'absence de pesanteur représente-t-il une attaque contre le père ? C'est tout à fait possible. Toujours est-il que dans les voyages interplanétaires, le ciel n'est plus au-dessus de nos têtes pour nous dominer. Nous sommes dans le ciel.

Au détachement de Gaia qui représente, comme cela a déjà été remarqué ici, l'éloignement de la Terre, je voudrais ajouter ce commentaire que cet éloignement comporte un rapprochement dans la mesure où il représente une maîtrise : celle que permet la vision globale depuis les satellites géostationnaires en particulier : par les satellites, la Terre perd son statut d'immensité inépuisable et donc mystérieuse et, par là, prend le statut d'un objet entièrement à portée de la main.

Pour revenir à l'absence de verticalité due à la non pesanteur, essayons de songer à ce que pourrait être la formation de l'inconscient des premiers enfants nés dans l'espace, c'est-à-dire en apesanteur.

Nous ne pouvons nous le représenter, car notre histoire terrienne nous barre l'accès à cette représentation. Il n'y a là pas moins un problème d'éthique aussi risqué que celui issu des techniques de procréation artificielle. En effet, nous sommes là en face d'une possibilité de manipulation phsychique qui ne dit pas son nom. Je sais bien que, biologiquement parlant, on n'a pas encore réussi à pratiquer l'embryogénèse en apesanteur (bien que les Soviétiques s'y soient essayés) mais il faut être vigilant.

Jusqu'ici, il y avait un domaine confiné, la terre. Et voilà que nous allons vers le déconfinement puisque le prochain domaine qui s'ouvre à nous, c'est l'univers entier. En ce sens il n'y a plus de limite ; mais, si un esprit épris de liberté peut s'en réjouir, il faut aussi voir qu'il n'y a plus d'au-delà, d'horizon à dépasser. Sauf un horizon temporel imposé par la durée de la vie humaine. Pour atteindre des étoiles un peu lointaines il faudra plus qu'une vie humaine. Alors, ce seront des générations entières qui seront engagées dans ces voyages. Même si des solutions techniques comme la congélation sont trouvées, la notion de moi individuel perdra de sa pertinence.

Voilà un nouveau sujet de méditation pour le psychologue. Mais j'en parle nécessairement en termes naïfs à partir de ma condition de terrien. C'est ce que j'appelle notre condition préspatiale. Cette condition nous place d'une certaine manière en position de créateur des temps futurs, que ce soit sous sa modalité maternelle ou paternelle. Il y a là sans doute un réservoir d'émotion qui contient l'énergie nécessaire à des opérations comme le présent colloque et qui plus généralement soutient les entreprises spatiales.

Dans le film que nous avons vu hier, l'astronaute Kathy Sullivan parlait, à propos de Voyager, de "téléprésence". Comme en écho à une remarque faite ici même par C. Cesarsky, qui disait qu'au-delà des justifications scientifiques il y avait le désir d'être physiquement présent dans l'espace. Même si nous-même n'irons pas en orbite, les mécanismes d'identification, que je laisse le soin aux psychanalystes d'analyser, font que nous avons le sentiment d'être partiellement là où les astronautes sont.

Mais revenons au terme de téléprésence. La présence, dit Heidegger, est une ouverture, une clairière, une entrée en présence. Et voilà qu'avec le "télé", la technique vient gauchir cette méditation ; les heideggeriens devraient être aiguillonnés par cette intrusion de la technique dans leur domaine. Nous observons qu'ici c'est l'ontologie elle-même qui est à l'épreuve.

Enfin, il est clair que l'espace devient un enjeu politico-économique, comme l'agriculture par exemple. Dans cette configuration, l'espace est un élément extérieur par rapport à une politique dont le sujet est terrestre. Mais si les affaires spatiales s'amplifient suffisamment, le sujet même de la politique sera l'Espace lui-même, du moins dans l'Espace, de sorte que nous aurons une nouvelle lecture du terme politique spatiale : l'Espace sera la nouvelle arène en politique et "politique spatiale" sera alors homogène à "politique occidentale" ou "politique africaine".

Voilà donc quatre domaines, la philosophie critique, l'éthique, la psychanalyse, le politique, qui ont leur place, si j'ose dire, dans l'espace, et il y en a d'autres.

Ainsi donc, non seulement l'Espace sous son aspect technologique intègre toutes les techniques -il faut en effet faire appel à toutes ses ressources les plus variées pour lancer une fusée- mais il est aussi en position d'intégrer les sciences exactes et les humanités ou l'anthropologie, comme j'ai essayé de le montrer par l'accumulation des exemples que je viens de présenter.

Ainsi, l'Espace serait ce lieu de dialogue entre les sciences exactes et la philosophie, suffisamment riche pour ne pas se réduire à la simple épistémologie, et serait justifié le terme d'Anthropologie Spatiale.

Un double écosystème se met par là en place : celui où des flux matériels circulent grâce à l'astronautique dans tout le système solaire, l'<u>écosystème solaire</u> en somme, et celui où grâce aux réfractions multiples que permet le terme Espace, des circulations se font entre l'espace intérieur de l'esprit et l'espace extérieur de la technique [1].

Je me suis risqué ici à parler comme scientifique de domaines qui ne sont pas mon métier, au risque de légères banalités. Mais c'est pour encourager nos amis philosophes à s'avancer de leur côté un peu plus sur le terrain scientifique. C'est par de tels premiers pas que le dialogue peut prendre forme. Et si mes propos contiennent quelque vérité, ils dépassent mon opinion personnelle et peuvent rencontrer votre assentiment. Ainsi, je suggère que la fin de ce Colloque ne soit pas un terme mais un début. Autrement dit, nous pourrions garder un contact organisé, quelque chose comme une Société d'Anthropologie Spatiale.

Je terminerai en laissant cette suggestion sur son orbite, en souhaitant qu'elle ait des effets dans le réel.

Référence

[1] J. Schneider, L'écosystème Terre-Espace, à paraître

Interventions

<u>Ruth Scheps</u>

L'Espace comme lieu de dialogue non seulement entre les sciences et la philosophie mais aussi avec l'art. En effet, l'Espace représente d'ores et déjà davantage qu'une source d'inspiration pour certains plasticiens et musiciens. Nous assistons donc aux débuts de l'Art spatial.

François De Gandt

Le philosophe est attentif aux idéologies et aux représentations du monde.

Seule une frange très étroite de gens peuvent accepter la représentation nouvelle du monde qui viendra de la conquête spatiale.

Quand on étudie l'histoire des idées, avec des étudiants du Japon ou de l'Afrique, on voit que pour eux l'image nouvelle du cosmos unifié du 17e siècle est nouvelle et forte. Descartes a écrit en 1640 environ qu'il y a une matière une et identique dans l'univers entier. Cela n'est pas encore admis par tant d'esprits sur notre planète.

Notre culture a vécu si longtemps sur une valorisation des différences dans l'espace (cf. l'univers de Dante).

Un effet idéologique important, pour notre monde occidental au moins : nous pensons que notre monde est fini c'est-à-dire clos, sans rien à découvrir, et aussi terminé, c'est-à-dire proche de l'essoufflement. La perspective de la conquête spatiale est un merveilleux andidote à ce désespoir.

Guy Pignolet

Nous avons ouvert une frontière, celle de la biosphère, qui nous donne accès à l'écosystème solaire, mais la technique ne nous donne pas encore l'accès physique à "l'univers" intersidéral. Il existe entre les confins du système solaire et "le reste de l"Univers" un désert considérablement plus difficile à franchir que ne l'a été la sortie de la planète Terre.

DE L'ESPACE PENSE A L'ESPACE INVESTI : RECHERCHE PHENOMENOLOGIQUE SUR LA CONDITION SPATIONAUTIQUE

Jean Seidengart
Maître de Conférences en philosophie
Université Paris X Nanterre

La présente communication ne se présente pas comme la plupart des exposés philosophiques actuels : elle ne se limite pas à un simple bilan de recherches textuelles, philologiques, historiques ou scientifiques. Bien qu'elle se réfère à des textes importants et consacrés par la communauté philosophique, elle a surtout le souci de recourir à la méthode phénoménologique pour analyser quelques traits saillants de la condition spationautique considérée comme une nouvelle manière d'être au monde et de l'appréhender. L'importance de ce nouveau phénomène met, à notre avis, le philosophe en demeure d'en recueillir le sens et de confronter celui-ci aux acquis de la phénoménologie.

Enfin, avant d'entrer dans le vif du sujet, nous tenons à remercier tout particulièrement l'Astronaute Wubbo Ockels pour l'aide précieuse et irremplaçable qu'il nous a si obligeamment apportée et sans qui notre phénoménologie aurait été privée de phénomènes vécus.

1. Retour à l'immédiatité de notre ancrage terrestre dans le monde

Commençons par élucider les multiples significations de notre rapport au monde en remontant aux déterminations originaires qui constituent notre enracinement primordial dans le monde environnant en deçà de toute sédimentation théorique d'ordre scientifique. Pour retrouver ce rapport originaire au monde, il faut percevoir le temps en oubliant nos horloges, montres et chronomètres, il faut saisir aussi les régions de l'espace en laissant de côté la géométrie et la géographie ; il nous faut nous orienter sans boussole, sans gyroscope, et regarder notre environnement existentiel sans recourir (autant que possible, car c'est une expérience-limite) aux sédimentations culturelles déposées dans notre mémoire. Retournons brièvement dans le monde primordial qui semble s'originer à partir de notre corps propre et qui se constitue à l'aide de schèmes généraux opérant sur l'afflux de nos impressions vécues (sans l'interposition d'instruments techniques ou d'appareils conceptuels théoriques). Ce retour au monde perçu indépendamment de tout savoir, ou plutôt antérieur à tout savoir et à toute construction théorique nous reconduit aux données sensibles

J. Schneider and M. Léger-Orine (eds.), Frontiers and Space Conquest, 105–115.
© 1988 by Kluwer Academic Publishers.

de la perception et à la mise en forme linguistique de celle-ci au moyen de la langue maternelle. Langage et perception sont pour nous le niveau originaire indépassable (car ineffaçable) de notre ancrage dans le Monde. Tous deux résultent de notre condition ambiguë de corps sujet et de conscience incarnée. Langage et perception expriment l'entrelacement de notre conscience et de notre corps propre dans la constitution de tout sens.

L'homme est ce pour qui et ce par qui tout acquiert un sens. L'humain est le lieu du sens. Toutefois, ce sens n'est ni dans l'homme ni dans les choses : il est dans la manière d'être au monde de l'homme, dans l'ouverture au monde. Le sens c'est en quelque sorte une recréation du monde à partir de notre ouverture, de notre déhiscence au donné.

Tandis que je puis choisir des objets dans le monde, m'en approcher ou m'en écarter, je ne puis ni m'écarter de mon corps, ni du monde. Le corps et le monde ne relèvent pas de notre choix, même si nous sommes responsables du sens que nous leur donnons. Donc, le corps propre n'est pas plus un "objet" que le monde. Je ne puis pas saisir la totalité des aspects du monde, ni celle de mon corps propre. En outre, avant toute prise de conscience, de ma part, le monde et le corps sont déjà là : ce sont les dimensions de mon ouverture à l'existence.

On ne saurait dire d'un humain qu'il est dans le monde en entendant par là qu'il y est contenu, inclus, compris comme une pièce dans une collection ou comme un organe dans un organisme. Dire que l'homme est au monde et non pas simplement dans le monde, cela signifie que l'homme est ce à partir de quoi (désignation d'une origine) s'ouvre le monde. Corrélativement, le monde n'est pas le contenant universel de tous les lieux qui se contient lui-même (Aristote), mais ce à quoi l'homme est ouvert, c'est-à-dire notre ancrage dans la réalité. Il apparaît maintenant avec force que le monde ou l'Univers est hanté par l'humain dont il reçoit son sens, tandis que l'homme a besoin de lui pour extérioriser son intériorité, et se ressaisir à partir des marques effectives de son objectivation.

Entre l'homme et le monde il y a une relation de réciprocité : l'un et l'autre s'entr'expriment en quelque sorte (vision du monde et monde vu).

Mon corps propre n'occupe pas une portion d'espace, mais, malgré ses dimensions dérisoires, il habite le monde, c'est-à-dire hante le monde et assure sa prise sur lui : il se l'approprie et se familiarise avec lui en y séjournant. Tout se passe comme si l'existence humaine présupposait comme son propre fondement une totalité d'appartenance englobante : même dans le rêve, la rêverie, l'hallucination et le délire, il y a rapport à un monde. C'est ce que les phénoménologues ont diversement reconnu et pris comme point de départ de leurs analyses respectives.

A partir des liens que tisse notre corps propre avec le monde surgit le phénomène de l'horizon. Celui-ci n'a pas d'existence propre, autonome ou absolue, il est bien un phénomène au sens husserlien du terme : il est tout entier dû au mode d'apparaître des choses selon ma perspective égologique. Il n'y a pas d'horizon en soi, mais toujours pour un sujet percevant, corrélatif de sa perspective. Si je me déplace sur la Terre, l'horizon recule en fonction même de ce mouvement et ne cesse d'accompagner mes déplacements. Mon horizon vient toujours enclore (bien que provisoirement) l'ouverture du paysage. Mieux, c'est l'horizon et ce qui me permet d'organiser l'unité du paysage à l'intérieur de son confinement circulaire dont je suis immanquablement le centre. Au-delà de l'horizon est l'absence, à l'intérieur de son confinement la présence. Mais comme la ligne d'horizon circulaire laisse surgir de nouveaux paysages à mesure que changent mes perspectives, la relation entre la présence et l'absence change à son tour. Donc, on ne saurait réduire le monde à l'horizon, car le monde apparaît plutôt comme ce qui n'en finit pas de surgir par-delà tous les horizons possibles. En ce sens, on pourrait définir le monde à l'instar d'Husserl et de Maurice Merleau-Ponty comme l'"horizon des horizons" [1], si, toutefois cette formule garde encore un sens.

L'ouverture au monde s'opère au sein de la fermeture de mon point de vue mobile. L'ouverture du paysage n'a elle-même de sens que par le confinement de l'horizon qui n'est que la projection de ma propre finitude.

Cette finitude n'a une prise sur le monde que par un phénomène de projection de l'intérieur du moi vers l'extériorité du monde qui apparaît initialement dans la distinction de directions privilégiées dans l'espace. Celles-ci s'organisent à partir du corps propre, référence originaire, centre d'orientation absolu à partir duquel s'ordonne la totalité du monde perçu : la droite et la gauche ; le haut et le bas ; l'avant et l'arrière. Avant d'élucider ces déterminations spatiales, il faut repartir de leur point d'intersection : l'ici (c'est-à-dire l'endroit où se trouve le sujet percevant ou parlant). Pas d'ici objectif, en soi, mais toujours en fonction de celui qui parle ou perçoit. A la limite il est impossible que chaque moi ne soit pas un ici. Où que j'aille, l'ici est là où je suis. Mon ici est le là d'autrui, son ici est mon là. Si je vais là, ce sera mon nouvel ici.

Autour de l'ici et du là s'articulent le prochain et ce qui est au voisinage du centre de perspective et d'initiative propre et le lointain, par essence périphérique. Le passage de l'ici au là relève d'un mouvement centrifuge, tandis que du là à l'ici le mouvement est centripète. L'ici est toujours pour moi le centre de mon horizon perceptif. L'ici est lié au bas (ici bas) dans la mesure où le sol nous sert de support, quelle que soit notre posture, en raison de notre impuissance à nous arracher naturellement à la terre. Le bas est la marque de notre faiblesse et de notre impuissance : c'est le lieu où la pesanteur finit toujours par triompher de nos efforts. C'est toujours par rapport au bas que nous déterminons notre posture et notre équilibration. Même en position assise il nous faut lutter pour maintenir le tonus musculaire de notre nuque afin d'éviter que notre tête ne bascule vers l'avant. D'où la nuance de mépris qui affecte le bas : la bassesse est le fait de tout ce qui se relâche ou de la lâcheté. Le bas est le lieu de la chute, de la faillibilité, de la décadence, de la déception et de l'échec.

Le bas ne se confond pas avec la Terre qui s'offre plutôt comme notre support immobile, le lieu de notre repos, notre substrat permanent. Malgré sa mystérieuse opacité, la terre est le berceau de notre humanité, le cordon ombilical qui nous relie de façon privilégiée à l'Univers où nous nous définirons toujours comme des terriens. Quel que soit l'avenir de notre humanité, par-delà toute errance spatiale notre histoire nous renverra sans cesse à notre origine terrestre de terriens.

Le haut se définit pour nous comme la direction de notre redressement postural et se confond avec notre accession à la dignité d'homme. La verticalité et l'élévation, c'est pour nous le symbole de la transcendance et du dépassement, c'est la direction métaphorique vers laquelle nous "surmontons" la négativité des obstacles. A la limite, c'est aussi le sens de l'hybris humaine surtout lorsqu'il est question d'une hauteur inaccessible comme le Ciel où s'élancent en vain les lignes infinies de notre désir d'éternité . L'élévation est la dimension des monts sacrés et de leur pendant artificiel, les grands monuments ziggurats, pyramides, tours sacrées dont l'ascension nous permet de quitter graduellement le monde horizontal de l'expérience profane.

En opposition à l'opacité terrestre le Ciel est le séjour divin de la lumière (Zeus, Dius-piter ; Dius = dieu, jour ; piter = père) qui traverse l'air, les nuées, les eaux et les corps transparents sans les briser. Corrélativement, notre vision a quelque chose de céleste qui culmine dans la vision mystique, la révélation et l'illumination religieuse.

Notre monde phénoménal est compris entre la Terre, le Ciel et l'horizon. A ce niveau déjà apparaît un ordre universel de co-appartenance où viennent s'équilibrer les éléments, pesanteur terrestre, grâce aérienne du ciel, opacité du sol et transparence des espaces célestes, opposés selon un jeu de combinaisons systématiques.

La Terre est plate, malgré les accidents de terrain, et son épaisseur est indéterminée : elle est la scène ou se joue le drame d'une existence, dont nous sommes à la fois les acteurs et les spectateurs. Tandis que la profondeur insondable de la voûte céleste, inaccessible et hors de toute atteinte, s'offre à notre vision comme un pur spectacle indépassable qui englobe et contient l'ensemble de nos actions et de nos représentations.

La Terre apparaît comme enchâssée à l'intérieur d'un ordre total où vient s'équilibrer le jeu des puissances opposées. Cet ordre total se manifeste avant tout à travers l'alternance de phénomènes célestes périodiques, cycliques dont le retour inlassable (et prévisible intuitivement) constitue pour l'homme le paradigme de la régularité.

2. Le recouvrement du monde vécu par la conceptualisation scientitfique

Que ce soit en s'en réjouissant, comme Cassirer [2], Brunschvicg [3] et Bachelard [4] par exemple, ou bien en le déplorant comme Bergson [5], Husserl [6] et Merleau-Ponty [7], la connaissance scientifique n'a cessé depuis son émergence à la Renaissance, d'élaborer des appareils conceptuels pour atteindre à l'intelligibilité globale de la réalité physique. Cependant cette conceptualisation scientifique de la nature eut pour conséquence de recourir et, par là-même, d'oblitérer le sens de notre rapport vécu au monde en éliminant tout élément puisé dans la structure préréflechie de l'expérience humaine. Les sciences positives ne parvinrent à prendre leur essor, leur autonomie et leur ampleur qu'en renonçant (difficilement mais sûrement) aux données immédiates de notre présence originaire au monde. Selon Husserl, le point de départ de ce recouvrement, c'est Galilée qui opéra ce qu'il appelle une "substruction" [8] : c'est-à-dire la substitution des idéalités mathématiques aux formes empiriques, et des constructions idéalisatrices aux données naturelles qui a pour effet de subvertir notre rapport au réel. Husserl écrit en ce sens : "*Ainsi, dès Galilée commence la substitution d'une nature idéalisée à la nature préscientifique donnée dans l'intuition. (...) Pourtant c'est en ce monde de l'intuition que nous vivons, conformément à notre mode d'être, c'est-à-dire dans toute la chair de notre personne. (...) C'est là une remarque importante bien que fort triviale. Car c'est précisément cette trivialité qui est masquée par la science exacte, donc par cette substitution d'une activité méthodiquement idéalisante à ce qui est donné immédiatement comme la réalité, et donné avec une force, une persistance, une vérité, dont la nature est unique et insurmontable*" [9].

Husserl, qui était lui-même élève de Weierstrass et de Kronecker ne renie rien de la pensée scientifique, mais il cherche simplement l'origine de la "crise", qui "affecte l'humanité européenne" et la pensée scientifique, donc la philosophie par voie de conséquence. Par sa découverte <Entdeckung>, Galilée et ses successeurs ont recouvert <verdeckt> le monde de la vie <Lebenswelt>. La mathématisation impose à ce monde de la vie un vêtement d'idées et de symboles <Ideenkleid> qui le dissimule et le travestit <verkleidert> dans la mesure où nous le prenons pour l'être vrai [10].

Pour illustrer cette opposition, on peut comparer l'espace intuitif perçu dont il a été précédemment question à l'espace pur de la géométrie classique. Il est clair que l'espace perçu n'est ni homogène, ni continu, ni isotrope, ni infini : en effet, la distinction de région et de direction privilégiées absolues est centrée sur le corps propre et reste confinée à l'intérieur de l'horizon perceptif qui soude la voûte céleste élevée à l'opacité du socle terrestre. Certes, il est vrai que certains psychologues et philosophes [11] ont vu des ressemblances entre la "géométrie spontanée" de la perception et la géométrie positive en se fondant sur les analyses de Félix Klein [12] : notamment l'invariance des relations spatiales au sein d'un certain groupe de transformations (déplacements, inversions, variation des dimensions, etc...) Si l'on suit la pensée husserlienne, toute construction théorique, toute pratique, sont des activités, mais celles-ci ne sont possibles que si elles prennent appui sur un socle, un sol qui leur sert proprement de fondement originaire. Or ce sol n'est ni construit, ni posé, mais renvoie à la "structure générale de notre réceptivité" [13]. La passivité et la réceptivité du sujet sont originelles, elles sont plus fondamentales que l'acte de poser. L'acte thétique (de position) ne peut exister sans un fond préalable, ainsi la non-thématisation est première et rend possible toute position. Le corps propre et la Terre sont les éléments fondamentaux de notre passivité : ce ne sont jamais des objets pour moi, c'est le sol de notre expérience. Le corps-sujet et la Terre sont ce à partir de quoi toute expérience est possible, ils forment le noyau des signification prédonnées qui ouvre le champ originaire de notre rapport au monde thétique. Le corps propre et la Terre relèvent donc de ce que Husserl appelle : le pré-théorique (ce qui est donné avant toute construction conceptuelle), le pré-réflexif (ce qui est antérieur à toute réflexion), l'anté-prédicatif (ce qui précède tout jugement).

Toute la difficulté de l'entreprise du dernier Husserl (1929 à 1939) a résidé dans ce questionnement en retour sur ce monde-de-la-vie, car l'analyse des structures de l'expérience préréfléchie et de l'antéprédicatif se fait au moyen de la réflexion et du jugement. La reconduction à l'origine ou à l'originaire ne peut totalement s'effectuer au niveau du discours rationnel mais s'opère plutôt dans une exploration de la totalité inhérente au vivre.

Tandis que la vie même de la conscience va de l'irréfléchi à la réflexion, du non-thématisé à la conceptualisation, la démarche du dernier Husserl procède en remontant le courant de la conscience. C'est ce que résume très clairement Maurice Merleau-Ponty [14] : "*La science n'a pas et n'aura jamais le même sens d'être que le monde perçu pour la simple raison qu'elle en est une détermination ou une explication (...). Revenir aux choses mêmes, c'est revenir à ce monde avant la connaissance dont la connaissance parle toujours, et à l'égard duquel toute détermination scientifique est abstraite, signitive, dépendante, comme la géographie à l'égard du paysage où nous avons d'abord appris ce que c'est qu'une forêt, une prairie ou une rivière*".

3. Le renversement du schéma husserlien : l'exploration spatiale transforme radicalement notre rapport à la Terre et au monde.

Certes, la phénoménologie husserlienne a montré de façon décisive qu'avant toute théorie, toute conceptualisation et toute explication scientifiques, se trouve le monde vécu, le monde-de-la-vie (Lebenswelt). Mais si l'on veut bien thématiser le cas du vol spatial, il faut bien reconnaître que le schéma est sinon inversé, du moins tout autre. La technologie et la connaissance scientifique contemporaines ont ouvert la voie de la conquête spatiale. Celle-ci est même devenue le paradigme des entreprises majeures de notre époque. Cependant, l'astronautique fournit la possibilité matérielle à l'homme d'effectuer des vols extra-atmosphériques et de vaincre les problèmes posés par le milieu spatial. Bien évidemment, il a fallu maîtriser ces problèmes *avant* d'effectuer des vols habités. Autrement dit, le nouveau rapport à l'espace nécessite une profonde connaissance préalable des contraintes physiques, physiologiques, biologiques et psychologiques que l'astronautique doit vaincre pour rendre supportables aux astronautes les conditions de leur vol spatial. Ce n'est qu'une fois vaincues ces difficultés que l'homme a pu enfin *vivre* dans l'espace.

Cette fois, l'entrée effective dans le milieu spatial, le rapport vécu au cosmos est *second*, si du moins on considère que l'entraînement des astronautes n'est qu'une situation *simulée*. Les vécus de vol spatial sont la *contrepartie subjective* d'un conditionnement scientifico-technologique. A cela on peut objecter que sur *terre déjà*, les moyens technologiques divers dont disposent les hommes suscitent les phénomènes d'appareil, ce que le dernier Kant de l'*Opus Postumum* appelait des "phénomènes de phénomènes". Ces "phénomènes de phénomènes" prennent leur sens par rapport aux données premières fondatives, anteprédicatives, qui nous prédonnent le monde. De cette façon, il est possible de situer la conquête spatiale dans le sillage de la "substruction" qui recouvre, oblitère ou éclipse le socle originaire, la couche primitive de représentation. C'est d'ailleurs ce que pense Husserl dans le célèbre inédit de 1934 [15] où il envisage de proche en proche qu'un engin (voiture, wagon, astronef, bref, un artefact qui nous permette de nous déplacer) et devienne à son tour un "corps de base-relatif" (contrairement au corps de base absolu qu'est la Terre) par rapport auquel mon corps-sujet soit en repos relatif. Husserl envisage même le cas d'un "*vaisseau spatial*" <Raumschiff> véritable "arche-volante". Dans cette situation, les astres changeraient de sens et seraient perçus à leur tour comme des "chez-soi" possibles, des "arches secondaires" sans que, pour autant, la Terre soit destituée de son caractère privilégié de sol originaire où s'enracine notre humanité. Au cas où des hommes viendraient à naître à bord de vaisseaux spatiaux, c'est par le truchement de leur historicité, de leurs attaches historiques, que ceux-ci seraient liés de façon privilégiée à la Terre. Toutefois, il nous semble que la pratique effective des vols spatiaux nous ouvre des perspectives phénoménologiques assez différentes de celles qu'Husserl avait envisagées dans son manuscrit de 1934.

1. Nous voulons remarquer que loin d'opposer l'espace pensé à l'espace vécu, le vol spatial produit un *nouveau genre de vécus* qui n'est plus simplement opposable à la conceptualisation scientifico-technique. Il s'agit cette fois de comprendre que l'*expérience vécue de l'espace* exprime sur le plan charnel de notre subjectivité le sens que prennent pour nous les phénomènes prédits par la théorie et canalisés par la technologie. Il y a correspondance et corrélation entre les phénomènes prévus par la théorie, provoqués par la technologie

et effectivement éprouvés par les cosmonautes dans leur subjectivité.

2. Aux yeux du philosophe, il est important de souligner un fait beaucoup plus remarquable. En effet, la Révolution Copernicienne, opérée en pensée, laissait pourtant intacte la signification de nos vécus effectifs de terriens, toujours placés dans une perspective implicitement géo- et anthropocentrique. Avec la mise en orbite circumterrestre de vaisseaux spatiaux, le renversement copernicien n'apparaît plus seulement comme une révolution intellectuelle et scientifique : c'est désormais une révolution vécue effectivement. Le vécu astronautique rejoint le plan théorique copernicien qui planétarise la Terre en la lançant dans l'espace ; l'opposition perceptive Terre/Ciel perd sa pertinence et sa prégnance. Ce n'est plus qu'une ancienne façon de parler.

3. Le texte de Husserl de 1934 portait sur la "Weltkonstitution" d'un point de vue cinématique, mais il laissait pour compte l'aspect proprement dynamique qui est pourtant d'une importance capitale pour le comportement humain dans l'espace extraterrestre.

Tout d'abord, l'absence de pesanteur efface au plus profond de notre corps-sujet le sentiment du haut et du bas. Nos muscles antigravifiques ne sont plus sollicités et la disparition du tonus de soutien nous fait perdre les renseignements kinésthésiques sur l'exacte situation de nos membres les uns par rapport aux autres. Notre corps-sujet devient fantômatique car le maintien de la posture n'est plus lié au tonus musculaire. D'où une impression agréable de flotter librement qui se monnaye par un phénomène de désorientation où les repères implicites de la spatialité antéprédicative sont quasiment perdus.

En apesanteur, ce qui est vécu c'est l'homogénéité de l'espace où toutes les directions deviennent équivalentes sur le plan kinésthésique. Le seul moyen de se repérer est d'ordre extéroceptif : c'est la vue qui prend le relais de la proprioceptivité. Mon corps visible me fournit par le détour de l'extériorité les renseignements intimes que je recevais auparavant de l'intérieur de ma chair. En l'absence de pesanteur, la face cachée du corps sujet s'éclipse totalement. Or, comme la vue est un sens à distance, il y a une perte du contact immédiat avec le monde ambiant, devant la raréfaction des stimuli tactiles. En outre, l'extrême facilité de la locomotion et de l'exécution de mouvements acrobatiques, efface de mon corps les conduites d'équilibration, et détruit avec elles les axes de coordonnées du milieu existentiel terrestre. D'où un important décalage entre les différents aspects extéroceptifs et proprioceptifs du champ sensoriel qui ne se correspondent et ne s'entrexpriment plus avec autant d'acuité qu'en état de pesanteur. Cependant, l'enthousiasme que suscite l'absence de pesanteur est tel que, malgré les inconvénients qu'il présente pour les tâches des astronautes, il est comparable à un état de grâce dont la perte, lors du retour au champ gravitationnel terrestre, est perçue comme une véritable servitude.

Par ailleurs, quoique le champ visuel reste toujours limité, la disparition de l'horizon perceptif transforme la structure globale de notre rapport au monde. L'indétermination des ténèbres rend insondable la profondeur du ciel qui perd son apparence de voûte élevée et disparaît au profit d'un espace "homogène". Le Ciel bleu n'est plus que l'enveloppe atmosphérique du globe terrestre. Sans la muette communauté des étoiles qui criblent la nuit éternelle et sans la perception de la Terre, ou de la Lune, le regard du cosmonaute, incapable d'une quelconque accommodation, se perdrait dans un champ visuel vide.

Sur le plan égologique, l'immersion du moi dans l'espace produit l'impression d'une profonde solitude proche de la déréliction au sens métaphysique du terme. Le sentiment de ne pas appartenir au milieu spatial et d'assister au retrait de la Terre enclôt l'existence spationautique dans l'étroit confinement du vaisseau spatial que seules les télécommunications peuvent arracher à son total isolement.

Si l'on se place au point de vue kantien, "*une profonde solitude est sublime, mais elle inspire l'effroi*" [16]. La contemplation de l'espace cosmique suscite en nous le sentiment du sublime corrélatif de la conscience de notre insignifiance. Tout se passe comme si le moi perdait le sens de son existence lorsqu'il s'abîme dans la perception de l'immensité cosmique. L'astronaute n'est donc bien qu'un explorateur de l'espace et non un habitant de l'espace. L'exploration n'est pas seulement observation, regard, contemplation, mais conquête. La conquête est un mode d'appropriation qui présuppose la négativité des obstacles à vaincre. Tel est le sens actuel du rapport de l'homme à l'espace : non pas dans l'opposition statique de type pascalien d'où ressort une évidente "disproportion" [17] entre des termes hétérogènes, mais une relation dynamique, conquérante. Certes, l'espace n'est jamais conquis, c'est plutôt ce dans quoi il y a conquête. Du reste, c'est la raison pour laquelle la distance métrique à la Terre n'est guère importante. Ce qui compte ici, c'est le début d'une ère nouvelle, la sortie hors du berceau de l'humanité, le franchissement d'une limite naturellement imposée à notre condition.

Quelles que soient la durée et l'ampleur des vols futurs, c'est l'ouverture de l'errance extraterrestre de l'humanité. C'est l'auto-arrachement de l'homme à sa destination terrestre. Deux conséquences radicalement opposées découlent de la conquête spatiale et constituent sa propre dialectique interne, qu'aucun principe régulateur ne vient présentement surmonter.

a. Dans l'ordre théorique et perceptif, l'astronaute devient un véritable Cosmothéoros, selon la formule que nous empruntons à Huygens assistant directement dans l'espace au renversement copernicien des apparences géocentriques et géostatiques. Le défilement de la Terre permet enfin d'objectiver celle-ci comme corps céleste, comme dit Husserl : "*Nous comprenons ainsi sans plus l'homogénéisation en sorte que la Terre, elle même soit un corps sur lequel, accidentellement nous rampons*" [18]. Toutefois, cette compréhension est à la fois intellectuelle et vécue. La planétarisation de la Terre met enfin d'accord connaissance et perception, concept et intuition qui s'entrexpriment réciproquement dans un phénomène répétable et valable pour tout observateur situé dans l'espace cosmique.

b. D'autre part, l'objectivation de la Terre comme planète est corrélative d'un éloignement, d'un retrait de celle-ci, qui ne peut laisser l'astronaute indifférent en tant que terrien, et qui ne peut nullement coïncider avec le rapport originaire de sa chair à la Terre. D'ailleurs, à son retour l'astronaute voit sur rapport à la Terre et ses relations à autrui profondément transformés. Il a, en effet, éprouvé la précarité et la fragilité extrême du berceau de l'humanité, pour ne rien dire de ses dimensions insignifiantes ne serait-ce qu'à l'échelle du système solaire. Par un paradoxe axiologique, le réduit terrestre où l'humanité est actuellement confinée revêt une valeur d'autant plus grande qu'il est notre seul bien commun, notre unique point d'ancrage dans l'Univers, infiniment rare et de ce fait infiniment précieux.

Références

[1] Husserl, Erfahrung und Urteil, 1939, rééd. 1954, p. 23-42
M. Merleau-Ponty, Phénoménologie de la perception, 1945, rééd. 1971, IIe partie, III, p. 381
[2] Cassirer, Substance et fonction, 1910, tr. fr. 1977, Ie Partie, Ch. 4, p. 137-235
[3] L. Brunschvicg l'expérience humaine et la causalité physique, 1922
[4] Gaston Bachelard, Formation de l'esprit scientifique, 1938, p. 13-15
[5] Bergson, La pensée et le mouvant, VI, p. 223-227 (PUF)
[6] Husserl, Krisis, II, § 9b, tr. fr., 1976, p. 33-38
[7] Maurice Merlau-Ponty, Phénomènologie de la perception, 1945, Avant-propos, p. III
[8] Husserl, Krisis, tr. fr. p. 45
[9] Husserl, Krisis, tr. fr. p. 58-59
[10] Husserl, Krisis, tr. fr. p. 60
[11] (Piaget, Géométrie spontanée des enfants, PUF, Cassirer, PFS III, IIè p., Ch. 3, tr. fr. p. 182
[12] F. Klein Programme d'Erlangen, 1893
[13] Husserl, Expérience et Jugement, Ière sect., ch. I, tr. fr. 1970, P.U.F., p. 83
[14] M. Merleau Ponty, Ph. perception, avant-propos p. III

[15] Husserl, inédit de mai 1934, manuscrit D 17, édité par Marvin Farber in Philosophical Essays in Memory of E. Husserl, Cambridge, USA, 1940, p. 307-325, traduction française par Didier Franck, revue Philosphie n° 1, janv. 1984, p. 5-21
[16] Kant, Observations sur le sentiment du beau et du sublime, 1764, AK, II, 209, tr. fr. Pléiade t. 1, p. 454
[17] Pascal, Oeuvres Complètes, Seuil, Coll. l'Intégrale, p. 525-528, "Disproportion de l'homme"
[18] Husserl, manuscrit D 17, édition Farber, p. 320, tr. fr. Didier Franck, p. 17
"*Ohne weiteres wird die Homogenisierung von uns so verstanden, daß die Erde selbst ein Körper sei, auf dem wir zufällig herumkriechen*"

Intervention

René Bost

- Vous avez fait allusion au cours de votre présentation à la perte de sensibilité tactile au cours d'un vol spatial. En fait, on a observé au contraire un développement de la sensibilité tactile, notamment au niveau des pieds dont les cosmonautes se servent plus dans l'espace qu'au sol : de même que la position érigée de l'Homme a permis de libérer ses mains et développer sa sensibilité tactile, l'apesanteur, en libérant les pieds leur donne une sensibilité accrue ou tout au moins un usage nouveau non conforme au schéma terrestre.

- Vous avez montré que l'individu cosmonaute qui est allé dans l'espace devrait être transformé ou tout au moins différent dans son comportement et sa vision du monde. Malgré toute l'admiration que l'on peut avoir pour des individus courageux, je pense que la question qui se pose est plus de savoir dans quelle mesure l'Homme (espèce) va se trouver transformé dans sa pensée et son comportement perceptible que permet l'espace. Je ne suis pas sûr que l'expérience individuelle du cosmonaute qui est resté sur orbite terrestre à 300 km de la Terre soit fondamentalement différente de celle des premiers Hommes qui ont volé dans l'atmosphère à quelques dizaines de km à la sensation physiologique près (perte de la stimulation gravifique qui n'est pas due au fait d'être au-dessus de l'atmosphère mais aux conditions de chute libre des véhicules orbitaux).

Pour l'homme qui est allé sur la Lune, le problème est certainement différent car l'éloignement était considérable.

LA CONQUETE DE L'ESPACE OU LES NOUVELLES FRONTIERES DE L'INTER-DIT

Isabelle Rieusset
Département des Sciences de la Communication
Université de Caen, France

1. L'esprit pionnier : une mythologie traditionnelle de la frontière

Repousser indéfiniment les limites, telle semble être une des ambitions les plus tenaces de l'humanité, que supporte plus particulièrement aujourd'hui la conquête de cette "Nouvelle Frontière" qu'est l'Espace. Expression d'une volonté de puissance, il se pourrait cependant que cette ambition ne conduise l'homme qu'à être le jouet d'un miroir aux alouettes.

L'esprit pionnier du "toujours plus loin" requiert en effet une double conception de la limite qui confine à une duplicité plus ou moins inconsciente, sinon au paradoxe. D'une part l'invitation à aller toujours au-delà se nourrit de la croyance qu'aucune limite ne saurait résister indéfiniment. L'Homme ne reconnaît dès lors aucune limite absolument comme telle.

Mais d'autre part l'idée même de repousser indéfiniment les limites suppose qu'après chacun de leur franchissement en surgisse aussitôt une nouvelle encore à vaincre et ce à l'infini. L'on joue d'un côté sur le refus des hommes de toute limite pour les amener à dépasser une nouvelle frontière. Mais on leur signifie derechef qu'ils n'ont fait que repousser celle-ci. Leur geste ferait dès lors la preuve de l'existence d'une limite absolue, infiniment plus forte que l'Homme. Vu sous cet angle, l'esprit pionnier relève de l'entreprise de Sisyphe condamné à hisser sans cesse un rocher sur une pente, d'où il retombe inéluctablement. Répétition absurde d'une démarche sublime il est vrai.

Mais cette volonté effrénée n'a pas pour seul écueil d'être fondée sur un double langage. Elle ne comporte cette ambiguïté que pour se soutenir d'une conception de l'infini qui nous vient de la métaphysique antique et méconnaît les apports fondamentaux des mathématiques modernes telles qu'elles se sont développées plus précisément chez Cantor. Celui-ci s'est en effet opposé à la conception aristotélicienne selon laquelle un infini réalisé "simul in actu" ne pourrait être que Dieu ; le domaine humain ne pouvant prétendre qu'à un infini qui s'énumère "alterum post alterum" dans la linéarité d'une succession sans fin.

Aux yeux de Cantor cette dernière conception désigne peut-être l'indéfini ou l'indéterminé, mais ne saurait correspondre à l' "Infini proprement dit" qui suppose de pouvoir être réalisé en acte simultané-

J. Schneider and M. Léger-Orine (eds.), Frontiers and Space Conquest, 117–131.
© 1988 by Kluwer Academic Publishers.

ment, c'est-à-dire énuméré d'une façon achevée. Or pour pouvoir construire des nombres transfinis qui répondent à cette exigence, Cantor est obligé de créer un nombre ω qui est une limite radicale à l'égard des nombres entiers en ce sens qu'au lieu de se présenter comme le nombre immédiatement postérieur à l'énumération inachevable des entiers natures, ω n'a pas <u>un</u> prédécesseur. Dans l'ensemble des entiers naturels le plus grand nombre imaginable sera donc aussi éloigné de ω que le plus petit. S' "*il est permis d'imaginer le nombre* ω [...] *comme une* ***limite*** *vers laquelle tendent les nombres* $\underline{n}$" [1] ce n'est plus dans la perspective d'une approximation successive qui poserait cette limite comme très lointaine certes, mais fondamentalement accessible. La coupure reste radicale entre ω et l'ensemble qu'il délimite.

Ce détour par les mathématiques pour montrer :

- que l'aspiration d'infini qui anime l'esprit pionnier ne saurait mener nulle part tant qu'elle reste fondée sur une conception archaïque de la limite ;

- et que paradoxalement, se donner les moyens de réaliser humainement l'infini, c'est accepter, voire poser, l'existence de limites absolues et radicales.

De ce point de vue, la conquête de l'Espace est héritière non pas de la mythologie prométhéenne et pionnière, mais de la théorie de la relativité en tant qu'elle pose comme limite indépassable la vitesse de la lumière. J'examinerai dans la suite de mon propos en quoi le fait d'accepter les conséquences de cette limite absolue, au lieu de prétendre dépasser un jour la vitesse de la lumière comme on l'a fait de celle du son, est lié à une nouvelle conception de la frontière et de la limite comme interface, plus propre à faire évoluer la conquête de l'Espace que la mythologie traditionnelle des "Frontiers".

Mais avant d'analyser cette nouvelle conception de la frontière, il faut encore mettre à jour quelques uns des pièges qui sont toujours à l'oeuvre non seulement dans les imageries traditionnelles, mais de façon beaucoup plus grave dans des documents aussi officiels que les Space Settlement Papers dont la mission était d'envisager l'avenir du programme spatial des U.S.A.

L'ensemble de ce rapport s'inscrit sous le signe de l'héritage de l'esprit pionnier, "the frontier heritage", expression aux traductions multiples en français dans la mesure où le terme "frontier" désigne en américain non seulement la frontière mais surtout les territoires établis au-delà de la frontière par les pionniers mêmes qui les ont conquis.

L'analogie est dès lors posée entre la conquête de l'Ouest et la conquête de l'Espace à partir de la célèbre phrase de Kennedy "Space -The New Frontier". Or les dangers de cette analogie ne se limitent pas aux connotations "infinitistes" examinées précédemment [2]. Ses auteurs en sont du reste conscients puisqu'ils déclarent : "*Yet the Space Settlement Working Group was conscious of the fact that the frontier experience was shared by much of the world, that the people who settled our frontier came from many lands and many cultures*" [3]. Mais cette concession aux ancêtres européens ne suffit pas à masquer la couleur radicalement nationaliste de cette référence et la fierté à se poser comme dignes successeurs de héros perçus comme essentiellement américains. Prendre l'analogie entre la conquête de l'Ouest et la conquête de l'Espace presqu'à la lettre, comme le fait ce rapport, ne nous amène-t-il pas à nous demander si les futurs descendants des pionniers de l'Espace n'envisageraient pas dans quelques années leurs ancêtres américains avec la même distance que les Américains ne le font parfois à l'égard des Européens ?

Observation que je soulève non pour réveiller des querelles dommageables et destinées à s'effacer dans le cadre de la conquête de l'Espace espérons-le, mais pour mettre en valeur plus que ne le font les Space Settlement Papers les velléités d'autonomie et de puissance qui pourraient s'emparer des futurs pionniers de l'Espace s'ils devaient se comporter come les pionniers de l'Ouest.

Cependant l'espoir de revivre ce grand mythe américain risque d'être déçu dès l'absence d'Indiens constatée. Ceux-ci se féliciteraient de ne pas encore être présents dans l'Espace. On répète qu'ils se racontent à mi-mot : "*Il paraît que la lune est un grand désert froid battu par les vents*". Mais un Indien plus averti répond toujours alors : "*C'est impossible, sinon ça fait longtemps qu'ils en auraient fait une réserve indienne*". Si les nouvelles terres de l'Espace sont au contraire difficiles mais intéressantes à mettre en valeur, ce ne sont plus les réserves indiennes qu'on risque de voir s'y établir mais les camps de forçats ou les nouveaux goulags, dans une tradition déjà internationale. Fort heureusement la compétence technique et scientifique, l'entraînement et la motivation nécessaires aux pionniers de l'Espace suffisent pour l'heure à nous préserver de toute forme de colonisation par des prisonniers.

Ces différents modèles carcéraux d'établissement dans l'Espace sont en effet non seulement peu souhaitables à suivre, mais de toute façon impossibles à réactualiser dans le cadre de la conquête de l'Espace. Celle-ci implique un éclatement du géocentrisme qui s'accommoderait mal d'un modèle colonisateur. Celui-ci suppose un contrôle centralisateur d'une métropole sur les colonies, même les plus reculées. La colonisation a beau créer un territoire national discontinu et hétérogène, elle prétend s'opposer à ce risque en renforçant la représentation d'un état uni autour d'un centre de décision et entouré d'une frontière conçue comme une clôture circulaire qui préserve hermétiquement l'unité nationale.

Ce modèle monocéphale soutient tous les totalitarismes militaires et en particulier le fascisme et le stalinisme à partir desquels G. Bataille en a analysé la structure [4]. Il pourrait peut-être inspirer une conquête guerrière de l'Espace, mais serait un frein considérable aux enjeux non seulement de son exploration mais de son habitation. Si comme l'ont analysé les fondateurs de l'Ecole Française de Sociologie, en particulier M. Mauss, et à leur suite G. Bataille, un véritable mythe ne se contente pas de nourrir l'imaginaire d'une communauté mais opère au coeur même du lien social [5], alors le type de communauté qu'il fonde dépendra de la structure symbolique de ce mythe. Si le mythe des conquérants de l'Espace est monocéphale, c'est-à-dire fondé sur la notion d'un seul chef (au double sens de tête et de dirigeant), d'un seul centre, d'un seul Dieu, autour duquel se fonde une communauté fermée entourée de sa frontière comme d'une clôture, et ne repoussant celle-ci que pour des guerres d'expansion, il est clair que ces nouvelles colonies seront non seulement militaires mais fascisantes ou en tout cas totalitaires. Si par contre les pionniers de l'Espace sont inspirés par le mythe d'Acéphale que G. Bataille propose comme alternative [6] et qui figure à ses yeux le choix lucide du décentrement, et l'assurance corollaire que le corps social ne soit plus subordonné à la tête, alors le nouveau type de lien social des communautés spatiales risque de nous surprendre et de nous apprendre.

Néanmoins les connotations telluriques [7] du mythe d'Acéphale le rendent impropre à supporter tous les enjeux de la conquête de l'Espace. Les Space Settlement Papers souffrent eux aussi d'une mythologie trop terrienne. Lorsqu'ils tentent d'envisager les modalités de la constitution de ce nouveau lien social de l'Espace, c'est à travers les images du pionnier rêvant de se fixer sur une terre (les fameuses <u>frontiers</u>, qui sont ces nouvelles terres au-delà de la frontière) et d'y établir ses racines. Non seulement toute l'imagerie agricole et tellurique se voit réinvestie dans la conquête de l'Espace, mais les connotations les plus régressives de ces mythologies terriennes tiennent aux valeurs d'immobilité et de fixation qui leur sont attachées, et semblent bien impropres à soutenir l'effort et l'imagination d'hommes qui vont être confrontés à la maîtrise du transport des personnes, des marchandieses et des informations dans un mouvement de communication permanent.

Il est peut-être temps de sortir du naturalisme tellurique, c'est-à-dire non seulement d'une phase de croyance par trop déterministe de l'influence de la planète Terre sur ses habitants (le milieu déterminant de l'homme étant le langage par lequel il transforme la nature en culture) mais aussi d'un rôle déterminant de la terre au sens de sol sur des êtres qui en état d'apesanteur ne sauraient éprouver le même attrait de fixation.

A l'opposé de ces mythologies terriennes du pionnier ou d'Acéphale, la figure mythologique de Hermès aux pieds ailés, dieu du mouvement, de l'échange et de la communication, dont le milieu naturel est l'espace, semble plus propre à représenter l'enjeu non seulement européen mais mondial de la conquête de l'Espace. A travers ce symbole de l'échange et de la circulation nous allons voir se profiler une toute autre conception de la frontière que celle traditionnellement léguée par la légende des pionniers de l'Ouest.

2. Une nouvelle conception de la frontière : Hermès ou le mythe de la communication

Au paradigme du "settlement" qui suggère les connotations de fixité, d'enracinement, voire de lourdeur terrienne, la figure d'Hermès oppose la légèreté et la vitesse due à ses pieds ailés ou à l'usage de son char présent dans certaines de ses apparitions mythologiques. Maître des transports aériens mais aussi de tout ce qui circule, il devient le Dieu de tout ce qui s'échange, c'est-à-dire à la fois les marchandises (il préside au commerce) mais surtout les informations. C'est à ce titre qu'il assure sa fonction essentielle : celle de messager des Dieux. Or quel est le rôle d'un messager s'il veut accomplir totalement sa mission ? C'est non seulement partir remettre le message dont on l'a chargé à son destinataire mais aussi revenir accompagné de la réponse de celui-ci. La fonction d'information et de communication n'est établie qu'à partir ce de va et vient. A ce titre le nouveau messager de l'Espace c'est la navette. Le drame de Challenger ne retire rien au formidable progrès qualitatif que représente la mise en oeuvre d'une navette, dont la mise au point reste d'autant plus nécessaire que par ce mouvement de va-et-vient, enfin possible avec l'Espace elle ouvre le champ non seulement à une vraie communication mais aussi à un "settlement" qui ne soit plus la réactualisation maladroite d'un événement de l'histoire terrienne. En effet si le discours de propagande qui auréole le programme spatial américain reste tributaire d'imageries mal adaptées aux enjeux de communication de l'Espace, son programme, enjeu bien plus essentiel, est au contraire au coeur d'un mythe plus propre encore que celui d'Hermès, parce que dégagé de toute imagerie inutile, à supporter l'actualisation d'une communauté intersidérale fondée sur l'échange et la communication. Ce mythe, c'est tout simplement la navette.

Si la navette n'est pas comme le pionnier ou Hermès une figure mythologique, et ne semble pas pouvoir nourrir une même fascination imaginaire, elle est un mythe au sens bien plus fondamental du terme, c'est-à-dire au sens ethnologique qui désigne les mythes vivants, tels qu'ils sont actualisés au coeur du tissu social dans des groupes où ils fondent la règle de la circulation des personnes et des biens. En ce sens la navette est l'exemple même de la fonction fondamentale du mythe : celle de tisser le lien social. Les premiers tisserands qui choisirent un mot dérivé de nef pour désigner cet instrument avaient déjà compris que vaisseaux et instruments de tissage avaient fondamentalement la même fonction symbolique : permettre par un mouvemant de va-et-vient de former un tissu de relations solides.

Cette conception du lien social comme tissu suppose la primauté donnée à l'échange et la communication. Elle s'oppose de ce fait à l'image d'un consensus social fondé sur l'appartenance à un territoire isolé de l'extérieur par la clôture hermétique d'une frontière. Le lien social prétend alors reposer non plus sur l'échange mais sur la protection contre tout échange. Toute communication avec le dehors apparaît comme une menace à l'intégrité d'un territoire qui seul assure l'homogénéité du lien social. Mais cette conception montre vite ses lacunes car elle ne permet pas d'assurer la circulation harmonieuse des personnes et des biens au sein même de ce territoire. Pour renforcer le lien social défaillant il faut donc dans cette logique donner l'illusion aux éléments du corps social d'être unis en s'opposant à tous ceux qui se situent au-delà de la clôture qui délimite leur appartenance commune à un groupe. Ce qui se situe au-delà de la frontière ne peut dans ce sens qu'être qualifié d'ennemi. Telle une ligne Maginot, un rideau de fer ou le bouclier anti-missiles que projette R. Reagan, la frontière est alors une ligne de défense, un rempart qui tente de s'opposer à toute communication. La navette au contraire suppose une conception de la frontière comme un lieu d'échange et de va-et-vient entre deux milieux mis en contact : ce qui n'est autre qu'une interface.

Pour que l'espace devienne véritablement cette Nouvelle Frontière nécessaire au monde post-moderne où les structures sociales, politiques, économiques et militaires sont transformées par cette nouvelle révolution culturelle qu'est l'apparition des nouvelles technologies de la communication, il faut donc qu'il fonctionne comme une interface. Ce qui suppose, comme l'analyse avec intelligence P. Virilio dans son ouvrage l'Espace Critique, "*une mutation atteignant la notion de limitation. La limitation, devient commutation*" [8]. "*Le protocole d'accès de la télématique succède alors à celui du portail*" [9]. "*La "grille des programmes" succédant aux grillages des clôtures*" [10]. Le fonctionnement d'une limite comme commutateur implique en effet une toute autre conception symbolique de l'espace dedans/dehors, fermé/ouvert. Ce n'est plus la frontière clôture qui en fermant hermétiquement un territoire s'oppose à toute circulation. C'est au contraire l'ouverture de la limite/interrupteur qui coupe la circulation, ou la rétablit en fermant le circuit. "*Ce qui était jusqu'à présent la frontière d'une matière, le "terminal" d'un matériau devient une voie d'accès dissimulée dans la plus imperceptible entité*" [11].

La frontière ou la limite n'est plus l'ultime ligne d'une surface. Elle n'est plus un confin mais un intervalle, un espace de médiation. Elle ne se situe plus au bout de mais entre. Ainsi l'Espace est-il à considérer comme Nouvelle Frontière non pas parce qu'il nous mènerait toujours plus loin, aux confins extrêmes de l'univers, comme le suppose l'analogie pionnière. Il est de notre Nouvelle Frontière au sens où il actualise une nouvelle conception de la frontière comme ce qui se situe entre, inter, et donc comme interface, comme intervalle, lieu de médiation et d'échange qui met en relation des points de notre globe par des satellites notamment. Nouvelle Frontière qui au lieu de séparer relie, qui au lieu de dresser la barrière d'interdits met en oeuvre l'espace de communication d'un inter-dire, qui fait que le plus lointain peut devenir proche et contigu grâce à l'interface qui bouleverse nos représentations topologiques en mettant en oeuvre des trajets pour lesquels la ligne droite n'est plus le plus court chemin d'un point à un autre. La vitesse de l'information par l'intermédiaire de l'Espace égale à celle de la lumière nous rend proches et instantanés des événements produits à de très grandes distances. Autrement dit au lieu de tenter de dépasser la limite de la vitesse de la lumière comme si on voulait s'en affranchir, on se sert au contraire de cette limite pour créer un espace d'interrelation simultanée, une frontière qui soit une interface et non plus une barrière. Au lieu de vouloir transgresser une limite absolue conçue comme un interdit, on transforme une limite irréductible en un inter-dire.

La quatrième dimension court-circuitée permet par l'instantanéité de l'information de dépasser les distances de l'espace euclidien. Mais si le temps semble vaincu sous sa forme physique, il réapparaît sous la forme sociale du "Time is money" qui se révèle peut-être comme notre véritable quatrième dimension. Chacun a fait l'expérience à la fois de la proximité d'une voix au téléphone alors que son destinataire est à des milliers de kilomètres, mais du coût plus ou moins exorbitant de l'opération. La communication par satellites rencontrant a fortiori cette barrière de la rentabilité.

Hermès préside bien à cette ère spatiale de communication et d'échange. Mais dieu du commerce, il symbolise aussi une ère où tout échange a valeur économique, et profite à ceux qui inventant des nouvelles frontières au fonctionnement traditionnel réinstallent des barrières et des interdits monnayables pour céder l'accès au passage et à l'interface.

3. De la mystification de l'interdit à la mise en oeuvre de l'inter-dire

Tout comme l'esprit pionnier relevait on l'a vu d'un double fonctionnement de la limite, la reconnaissance de la frontière ou de la limite comme lien privilégié d'échange s'accompagne souvent aussi d'une stratégie duplice.

Alors que le fonctionnement de la frontière comme interface est fondamentalement disponible à la libre communication, son contrôle peut fort bien réserver cet usage à quelques initiés capables dès lors d'utiliser ce même dispositif à des fins différentes voire contraires. On a vu précédemment que la limite comme interface renversait l'assimilation du système fermé avec la fin de toute communication. De fait les techniques de la police de l'air et des frontières qui seront peut-être plus tard celles de l'Espace ont su se moderniser et transformer la stratégie traditionnelle qui consistait à cerner le suspect dans son repaire pour tenter de la piéger dans ses lieux de circulation obligés. "*Dès lors, il ne s'agissait plus, comme par le passé, d'isoler par l'enfermement le contagieux ou le suspect, il s'agissait surtout de l'intercepter sur son trajet, le temps d'ausculter son vêtement, son bagage, d'où cette soudaine prolifération de caméras, de radars et de détecteurs dans les lieux de passage obligés*" [12]. Mais ces équipements utilisés dans les aérogares allaient aussi servir dans les quartiers de haute sécurité dont la topologie dedans/dehors et la conception de la frontière qui les sépare sont loin d'être celles d'une interface ; Ainsi "*l'équipement de la plus grande liberté de déplacement* [servait] *paradoxalement de modèle à celui de l'incarcération pénitentiaire*" [13]. Non que ce système fût de mauvais aloi. Mais il est un exemple incontestable de réversibilité de la maîtrise des frontières et des limites, selon la fin poursuivie.

Les stratégies militaires et commerciales risquent peut-être à ce titre d'entrer en contradiction sur le terrain des Nouvelles Frontières de l'Espace. Entre la volonté d'une libre circulation des hommes et des marchandises et les volontés militaires de suprématie des Grands, le contexte politique devra choisir. Mais sachons d'ores et déjà que les choix éthiques ne seront pas seulement entre des conceptions traditionnelles de la frontière comme le rempart représenté par exemple par le bouclier anti-missiles et une nouvelle frontière interface telle qu'elle est réalisée par les satellites. Les stratégies militaires sont déjà à l'heure des nouvelles frontières. Elles manipulent avec un savoir-faire certain les possiblités ouvertes par les interfaces. P. Virilio constate que "*L'interactivité des moyens de communication de la destruction rejoint donc maintenant celle des moyens d'information : l'instantanéité de l'interface télévisée devient celle de la visée, celle de la non séparabilité des événements militaires dans la conduite de tir des représailles qui sous-tend la dissuasion*" [14].

Si l'étymologie du mot frontière venue de l'ancien français "frontier-ère" signifiant "qui fait face à, voisin" a fait place à une conception "*d'une guerre "électronique" tout autant qu'"atomique" où l'interface instantanée succède au face à face des armées*" [15], son sens militaire originel dès le XIIe siècle, à savoir "front d'armée" et "place forte", lui reste manifestement attaché.

Néanmoins il est regrettable de rejeter comme P. Virilio dans un même geste théorie de la relativité, mécanique quantique et guerre atomique ou télématique moderne comme relevant d'une même démarche dont l'aboutissement logique serait la "*guerre pure*" [16] sous prétexte que l'instantanéité de l'interface conduirait à l'idée d'attaque pré-emptive et de là à la confusion des notions d'attaque et de défense. Si l'usage militaire des nouvelles technologies peut faire craindre une amplification démesurée des fureurs guerrières de l'homme, il me semble abusif de prétendre que l'interface est coupable de la confusion entre attaque et défense. Un certain Machiavel déjà ... et les enfants... répètent depuis si longtemps ce précepte bien connu des stratèges que la meilleure défense c'est l'attaque.

Ne pas confondre donc l'usage militaire de découvertes techniques ou scientifiques, du reste souvent issues pour cause de crédits de recherches militaires, avec les multiples possiblités offertes par ces découvertes. D'autant plus qu'un des intérêts de cette utilisation militaire des interfaces qui pourra sembler à certains mineur au regard des risques encourus, mais mérite néanmoins qu'on s'y arrête, est de manifester clairement aux yeux de tous que le pouvoir se joue dans la maîtrise des lieux d'échange et de circulation, qu'il s'agisse d'hommes, de biens, de monnaies, ou d'informations, autrement dit de langage. La polarisation militaire sur les interfaces montre que c'est bien dans l'inter-dire, dans cet entre où se nouent à la fois le lien social et le langage, que se jouent les véritables enjeux. Les militaires se contentent, forts de ce savoir, de tenter de contrôler la maîtrise de ces interfaces. Jusqu'à présent les groupes gardiens du pouvoir et du maintien de la loi avaient tenté de masquer l'existence même de cet enjeu. Non contents d'interdire l'accès matériel à ces interfaces, ils en occultaient non seulement l'enjeu mais l'existence. Ainsi l'interdit d'accès matériel se doublait d'un interdit d'accès conceptuel qui occultait l'inter-dire. Tel a été le fonctionnement des représentants des systèmes métaphysiques ou religieux.

La question s'est jouée dans ce domaine autour de la notion de limite, mais dans sa dimension concentuelle où elle figure la loi, et non plus dans sa dimension contingente où elle se concrétise comme frontière.

Je vais faire appel pour illustrer mon propos à un court texte de Kafka intitulé Devant la loi qui condense de façon saisissante ce fonctionnement métaphysique de la limite et de la loi. "*Devant la loi se dresse le gardien de la porte. Un homme de la campagne se présente et demande à entrer dans la loi. Mais le gardien dit que pour l'instant il ne peut pas lui accorder l'entrée*" [17]. Parce qu'il prétend se tenir en sentinelle devant la loi, le gardien de la porte met en place une topologie où la loi se situerait de l'autre côté d'une frontière, dont le passage est soumis à un interdit. Topologie qui est déjà en tant que telle un simulacre, car la loi n'est pas cachée derrière cette porte entrouverte puisqu'elle opère au lieu même de cette porte, en tant que son passage serait la transgression de la loi. Or ce passage fût-il muni d'une porte n'en requiert pas pour autant la détention d'une clef. C'est en effet ouverte que la porte interdit le passage [18]. En quoi elle fonctionne comme un interrupteur. Ouvert il interrompt la communication. Fermé il permet le passage de l'information dans le circuit. Le gardien de la loi le sait, lui qui s'efface devant la porte ouverte sans crainte de voir pénétrer le secret de la loi par l'homme de la campagne. Lui qui face à cet homme agonisant après une vie en arrêt devant cette porte lui signifiera à la fin du texte : "*Ici, nul autre que toi ne pouvait pénétrer, car cette entrée n'était faite que pour toi. Maintenant, je m'en vais, et je ferme la porte*" [19]. Mais pour profiter de cette entrée, il fallait comprendre qu'elle fonctionnait comme une interface, c'est-à-dire comme une limite conçue comme commutation et non pas comme un au-delà repoussé indéfiniment. Or le gardien de la loi entretient l'homme de la campagne dans le leurre de cette fausse conception de la limite et de la loi. Il lui déclare en effet : "*Si cela t'attire tellement*, dit-il, *essaie d'entrer malgré ma défense. Mais retiens ceci : je suis puissant. Et je ne suis que le dernier des gardiens. Devant chaque salle il y a des gardiens de plus en plus puissants, je ne puis même pas supporter l'aspect du troisième après moi*" [20].

Le gardien de la loi joue le rôle du tentateur qui invite à la transgression de l'interdit. Il sait en effet qu'il ne court en ce cas aucun risque. Que l'homme de la campagne franchisse ou non cette porte, qu'il cède au désir de la transgression ou à la peur de l'interdit, il restera persuadé que derrière cette porte il y a une autre salle avec un autre gardien puis une autre porte et ainsi à l'infini, et donc que le secret de la loi est pour lui inaccessible, si bien gardée elle est par ses représentants. Il se pliera donc à leur loi faute de pouvoir rencontrer la loi. Au contraire le gardien de la loi, qui tel Saint Pierre possède lui les clefs symboliques de la porte, sait que la loi n'est pas cachée à l'infini derrière ces salles, mais qu'elle est une interface dont il suffit de maîtriser le code, un lieu de passage dont il faut s'assurer la maîtrise, ce qu'il fait.

Ainsi agissent la plupart des représentants des systèmes métaphysiques ou religieux, qui sous prétexte d'être détenteurs du secret d'un au-delà, imposent leurs propres interdits, tout en se réservant la maîtrise des interfaces, des lieux de passage et de circulation, en particulier monétaires, pour asseoir leur pouvoir. Les "détenteurs" des clefs de Saint-Pierre vendirent en leur temps leurs indulgences qui ne sont autres que ce droit de passage pour la porte céleste. Bien avant la religion chrétienne, les métaphysiciens antiques jouaient déjà des charmes de l'au-delà, pour préserver le caractère inneffable et mystérieux de la limite et de la loi. "*Le récit de l'interdit est interdit*" [21]. L'interdit majeur de la loi, comme de la limite, consiste dans ce refoulement qui occulte sa propre production comme effet des interférences du dire, du langage. En interdisant le déchiffrement de sa génèse, la limite métaphysique se pose comme loi transcendante, absolue, éternelle. C'est de cette impossibilité à la replacer dans la chaîne des échanges dont elle s'est abstraite arbitrairement qu'elle prétend fonder son pouvoir et sa légitimité à pouvoir déterminer le cours des échanges auxquels elle n'aurait point de part en tant que telle. La limite et la loi nous viendraient toujours d'un ailleurs, d'un au-delà. Ses tables seraient données aux hommes par un Dieu, ou plutôt par ses représentants qui sont censés avoir reçu la révélation d'un secret. C'est un principe transcendant incarnant le Un, l'Absolu, qui fonde dès lors la limite et la loi, et non plus la relation de l'inter-dit. Au lieu d'admettre qu'elle est le produit de cet entre dire, on fait comme si elle était antérieure à toute relation. On abstrait l'interdit de ce tissu d'interférences et d'échanges où il prend forme pour en faire un principe absolu. Si l'on considère au contraire la loi comme une production de l'inter-dit, elle est immanente aux structures d'échanges qui relient des sujets par des frontières fonctionnant comme des interfaces.

4. Prospective éthique et conquête de l'Espace : du prédire à l'inter-dire

La conception de la limite et de la loi comme interdit métaphysique ou comme inter-dire engage en effet deux conceptions du rôle de l'éthique, qui ne sont pas sans conséquence pour la prospective de la conquête de l'Espace. Celle-ci exige qu'une nouvelle éthique soit fondée qui soit capable d'assumer la confrontation de l'inter-dire sans avoir besoin de la mystification des interdits.

Là où une conception métaphysique de la limite et de la loi envisage l'avenir comme ce domaine de l'au-delà que seuls quelques initiés au secret divin peuvent prédire, la prospective moderne exige un dire qui sans prétendre à une révélation transcendante interagisse avec le réel dans son devenir. En matière d'éthique, la conquête de l'Espace n'a pas besoin de prophètes mais d'interprètes. Les spécialistes du pré-dire qui manipulent les interdits divins au risque de tenter quelques apprentis sorciers doivent faire place aux spécialistes de l'inter-dire, à savoir non seulement de l'interprétation des différents dires rencontrés, des différents codes par lesquels nous allons communiquer avec et dans l'Espace, mais de l'interactivité des réseaux d'échange créés sur le devenir du lien social de la nouvelle communauté humaine interstellaire. Ainsi la prospective éthique ne doit pas être une futurologie entretenant la coupure entre un avenir devenu objet de prédilection rhétorique et des actes présents privés de tout langage. Elle doit s'inscrire au coeur de "*l'expérience, c'est-à-dire*, selon G. Bataille, *le passage de la connaissance à l'acte*" [22]. Elle doit être une interaction au sens littéral de ce qui "vient entre", de ce qui maintient l'ouverture d'une mise en question, l'espacement d'une interrogation au coeur même de l'action.

L'éthique ne peut plus se poser comme un méta-discours qui s'abstrait de la réalité qu'il juge pour prétendre à l'illusion d'une objectivité. Elle doit au contraire interférer avec l'action qu'elle interroge dans l'interactivité constante du dire et du faire.

Références :

[1] Cf. Georges Cantor, cité par J.L. Houdebine in l'Infini n° 4, Automne 1983, p. 100

[2] cf. The Space Settlement Papers, in Journal of the British Interplanetary Society, 39, p. 291 : "*we are still pushing at the Frontier.*" "*and push back the Frontier a little bit more*".

[3] The Space Settlement Papers, Ibid

[4] Cf. G. Bataille, Propositions sur le fascisme, in Acéphale, janvier 1937, Eds J.M. Place, p. 17

[5] Cf. I. Rieusset, Le collège de Sociologie : G. Bataille et la question du mythe, in Actes du Colloque Georges Bataille et les Ethnologies, Editions de la Maison des Sciences de l'Homme

[6] Cf. G. Bataille, Chronique nietzschéenne (1) in Acéphale, juillet 1937, Eds J.M. Place, p. 21-22 : "*A l'unité césarienne que fonde un chef, s'oppose la communauté sans chef liée par l'image obsédante d'une tragédie. [...] Chercher la communauté humaine SANS TETE est chercher la tragédie*".

[7] Cf. G. Bataille, En marge d'Acéphale, in O.C. tome 2, Gallimard, p. 278 : "*le soufre est une matière qui provient de l'intérieur de la terre et n'en sort que par la bouche des volcans. Cela a évidemment un sens en rapport avec le caractère chtonien de la réalité mythique que nous poursuivons. Cela a aussi un sens que les racines d'un arbre s'enfoncent profondément dans la terre*".

[8] Cf. P. Virilio, l'Espace critique, Christian Bourgois, Editeur, 1984, p. 18
[9] Ibid. p. 14
[10] Ibid. p. 20
[11] Ibid. p. 18
[12] L'espace critique, op.cit., p.11
[13] Ibid. p. 11
[14] Ibid. p. 168
[15] Ibid. p. 177
[16] Ibid. p. 169
[17] Cf. F. Kafka, Devant la loi, in La faculté de Juger, Editions de Minuit, 1985, p. 100
[18] Cf. Ibid. p. 100 : "*Le gardien s'efface devant la porte, ouverte comme toujours*".
[19] Ibid. p. 101
[20] Ibid. p. 100
[21] Cf. J. Derrida,Préjugés devant la loi, in La Faculté de juger, op. cit., p. 118
[22] Cf. G. Bataille, Texte recopié par H. Dubief, Bibl# C'est la terre à découvrir, à conquérir, à exploiter. Ce n'est pas une barrière. Au contraire, c'est un vide à occuper. L'Alaska a été baptisée "the last frontier" (la dernière terre vierge). L'administration reaganienne, composée d'hommes de l'Ouest, souhaite réveiller "l'esprit pionnier", l'esprit d'aventure et de conquête. L'Espace est appelé "the new frontier". C'est une frontière "illimitée".

Interventions

Emilio Ortiz

Il faut prendre garde à ne pas se méprendre sur la signification que les Américains donnent au mot "frontier" qu'on voit souvent traduit ou interprété dans le sens de "ligne frontière", de "limite". Pour ceci, les Américains utilisent le mot "boundary". The "frontier", c'est tout autre chose : lors de la poussée vers l'Ouest, on désignait ainsi la zone mal définie bordant les terres non encore explorées. Il ne s'agit donc pas d'une ligne, mais d'une surface (comme pouvait l'être "le front" en 1914-1918). Par extension, "frontier" désigne la terre qui est devant, "the land that's in front". C'est la terre à découvrir, à conquérir, à exploiter. Ce n'est pas une barrière. Au contraire, c'est un vide à occuper. L'Alaska a été baptisée "the last frontier" (la dernière terre vierge). L'administration reaganienne, composée d'hommes de l'Ouest, souhaite réveiller "l'esprit pionnier", l'esprit d'aventure et de conquête. L'Espace est appelé "the new frontier". C'est une frontière "illimitée".

Robert R. Brownlee

First of all, permit me to say how important I believe this meeting to be, for it recognizes a prominent need to mesh technology with various existing philosophies, the better to serve all mankind. I thank ou hosts for making this meeting possible, and for creating an opportunity for a kind of non-tutorial communication. This is an excellent beginning to what I hope will be a continuing effort.

From time to time since we began 3 days ago, references have been made to the settlement of the New World. I gain the impression that most of you here have concluded that there is little in this experience that is relevant to the settlement of space. But these are aspects of the so-called frontier experience that seem to me to the waven into the space programs of the United States and I therefore believe relevance does exist.

I hope you can forgive me for speaking very personally on this point. I must do so for two reasons. Firstly, I am not, nor have I been a part of the U.S. manned space program. I am not in a position to speak for others in any way. Secondly, I am a creature of my own experiences, and can only bring them to bear on any new subject. I am aware, However, from my association with colleagues across the U.S. and from our literature, that I share view with many Americans.

A bit over a century ago, each of my 4 grandparents came to the western part of the United States via either a wagon or a horse. I therefore grew up in an atmosphere steeped in the philosophy of pioneers, and in an environment that was and still is in the process of being bent to the while of the descendants of those pioneers. Models that perhaps existed for this process were unknown by my grandparents and parents and life's pragmatic requirements were dealt with as they arose.

I feel very strongly that it was the attitute of the pioneers to the world around them that has survived in America, and that has been reponsible for helping to shape many of our activities, including our space programs.

Please make sure you recognize that the pioneering experience in Canada and the United States is not just 200 years old, or 100 years old. Even as i speak immigrants, both legal and illegal, are arriving in these countries in large numbers. Their diversities in race, nationalities and skills are so broad that we are having difficulties in their assimilation. But we all know, almost intuitively, why they are coming. It is not just for food, though that is part of the truth, nor is it just for jobs, and security. We are still too close to the dreams and visions of our pioneers ancestors to be scornful of theirs. Indeed, these people serve to renew the legends of the American West, to keep the old dreams alive, to make up-dates to our visions, and to help us make them come true.

On Wednesday, Madame Rieusset, if I understood here correctly, suggested that just as Americans had cut her ties to Europe, Americans in space might desire to cut their ties with hearth. I think you should know that we expect them to do so. We actually want them to do so! If I were still alive I would want to help them to do so! We have no real desire to export into space governmental controls, or perhaps even worse, govermental bureaucracy. Obviously that's the way we're starting out, but even now many of us find the necessity abhorrant. We'll be trying to reduce if whenever we find ways to do as.

Professor van de Hulst has reminded us that the only step we must take is the next one. He said nothing about how to determine the direction of that step.

For many Americans, they've experienced, for the most part successfully, choosing the direction of that next step by intuition. They step in the direction of their dreams and visions, of their hopes, without too much concern for current of future obstacles. They step with the knowledge that they've made mistakes in the past, and that they will err again. They still move, sometimes at least, ignoring available models, or without models, with high confidence that mistakes can be corrected, that new tools will come to hand, enabling them to overcome or to circomvent any obstacle. Americans tend to play down existing realtivies, I thin, even as they are occasionally pointed out to us by our European friends, in the expectation that new realities will be discovered as needed.

Let me say, as an aside, that I am surprised not to have heard here a spirited discussion of the ramifications to philosophers and to all mankind of genetic engineering. It is obvious that there are some who expect this new tool to enable us to evolve in a twinkling to a selected adaptation to whatever space environment we choose to like. Surely when life begins to choose what it will become, the subject deserves the most intensive study and discussion!

But returning to my main point, it seems to me that the American attitude I've just described in the same old pioneer attitude I first came to know as a long. This attitude is alive and well in the New World, ans it runs through our space program like a mighty stream, if occasionaly through a desolate terrain. I think that much our behaviour can be said to be derived from our optimistic philosophy. This philosophy clearly contrasts sharply with the traditional continental European philosophy, one that I have seen in several different ways at this meeting. How we need to rise above ourselves and our native philosophies if we are to achieve our maximum potential in Space!

An additional point. Space programs are ongoing in a number of places around the world. They will proceed, even though their rates are uncertain, unpredictable. These programs tend to enlist the visionaries, the optimists. Few participants are recuited from the ranks of the pessimistic. This will be true for people, not jus Americans.

Yet pessismists play an exceedingly important role in any endeavor. They tag the optimists with realism. Several times at this meeting I have heard new ideas about problems we are going to encounter, and the nasty ramifications to a bad decision. But for the nast part, these remarks seem to me to have had the flovour of "Someone should look into this", or "Someone ought to be careful there". The someone implicitly referred to seems to be someone else. If you have concerns, then you should do something about them. Someone else may do it too late, or not at all.

A world filled with space technologies desperately needs philosophers. But philospheres will have to decide for themselves to enlist in existing programs. If they do not, I believe then that everybody a space program will proceed without them, to be detriment of all mankind.

SETTLING SPACE
(THE COSMIC EXPANSION OF THE BIOSPHERE)

Eric M. Jones
Earth and Space Science Division
Los Alamos National Laboratory
Los Alamos NM 87544 USA
&
Department of History
University of Alaska-Anchorage
Anchorage AK 99508 USA

We are going settle space.

The first explorers -20th Century equivalents of Columbus and Cabot, Cartier and Bering- have already returned to port with reports of new continents. Advance parties will soon return to the Moon, go to Mars, and travel among the asteroids to make preparations for settlers who will follow: settlers who will build homes, raise families, and find ways to be self-supporting, much as terrestrial pioneers have done in the past.

Or, at least, that is what I expect.

Our journey into space began in the minds of visionaries: Tsiolkovsky, Goddard, Obereth, and others, who described cities in space and flights to the Moon and planets. The first realities were the flights of frails, temperamental rockets; these were humble beginnings indeed. But we made progress and stepped a bit closer toward making the larger visions into realities. Sputnik. Gararin's orbital flight. Communications satellites. The lunar landings...

As Arthur C. Clarke proudly notes in his book the Promise of Space, the visionaries had clear sight. I recently re-read Clarke's books from the 1950s -the ones with the lovely drawings by R.A. Smith. Looking at a thirty-year-old concept of a free-flying spaceman and comparing it with a photograph of Bruce McCandless flying the Manned Maneuvering Unit, we see the differences in detail. But overall, the vision has only needed a little fine-tuning.

The fidelity of our creations arises in part because thus far we have been doing the simple things : learning how to move about and to function in this novel environment. Our spacemen look and act like the explorers they are. The facilities they will build first on the lunar surface will be explorer's quarters, with a familiar, "polar" look. To this stage of the game we have been dealing with many of the same challenges that the early polar explorers faced, and so it should not be surprising that the costumes and the facilities are familiar.

J. Schneider and M. Léger-Orine (eds.), Frontiers and Space Conquest, 133–139.
© 1988 by Kluwer Academic Publishers.

I think that as we move into the settlement phase things will seem unfamiliar again for a while. The similarity to frontiers of the past will become more apparent as we gain experience with space settlement. But, for the present, we use words like "settlement" and "frontier" and "pioneers" and hope that others share our faith that like the frontiers of the past, the Moon, Mars, the Lagrange points, and the asteroid belt will not be wilderness forever.

Yet, when we look at Smith's drawings or the Apollo photographs, we see spacemen standing in the midst of what Buzz Aldrin has called "magnificent desolation". The Moon does not look like the sort of place where people live; there are no trees nor clouds, no grass, no water -only grey sand, a black sky, and a harsh sun. If the Moon visibly resembles anywhere on Earth, it is the deserts where virtually no one lives -the Sahara or the South Polar Plateau. It takes a great leap of faith and daring to evoke a vision of settlement in the misdt of this difficult reality.

For those of us who have been caught up in the vision since childhood, this leap of faith is a small one. The spirit of Robinson Crusoe leads us onward. Armed with some knowledge of Solar System resources, we believe that determination and inventiveness -the old pioneering spirit- will bring order out of chaos, settlements out of the wilderness.

Looking ahead for a moment, let me make the rather unstartling prediction that we will concentrate our efforts on the resource-rich places.

We will return to the Moon for several reasons. First, and certainly most important, because it is there and visible -the symbolic high ground. Second, we will return to the Moon because it is a training ground where we can learn some basic skills -how living things develop in fractional gravity, how to use poor ores and build relatively inexpensive shelter out of native materials, how to balance private, national, and international interests, and so on. Third, the Moon is likely to be an important source of building materials and propellants needed for orbital facilities and deep-space missions.

We will reach out for earth-approaching asteroids because they are rich in hydrogen, carbon, and nitrogen which are scarce on the Moon, and mine them for useful metals. Although fewer than a hundred are now known, like the Moon, earth-approaching asteroids are relatively accessible and will support the early stages of development of the inner Solar System.

We will reach for the martian moons -captured asteroids we think- and hence possible sources of supply for lunar operations, and surely vital to our presence on Mars.

We will reach down to the surface of planet Mercury and, using lunar-learned technologies like mass drivers and orbital smelters, build big power stations to tap the enormous and dependable fusion reactor that nature has given us. Solar energy, beamed as microwaves or laser light, may push the thin sails of craft bound for communities among the main-belt asteroids -those portable mountains that may be the true material treasures of the solar system- or even toward the comet cloud far beyond the orbits of the outer planets. Solar energy for transport and development.

But what kind of communities will we create on the Moon or Mars, in orbit or among the asteroid and comet swarms, and how will they live ? There are no trees. No fields, No clouds.

With regard to the long term, our vision cannot be clear ; all we can say with confidence is that the reality will be different in some ways from anything we can imagine. In the near term, the reality will, of course, be very close to what we do imagine. We will try to re-create what we know. The new patterns will only emerge with experience and time.

Let me sketch a terrestrial example.

There is a part of Australia called the Mallee in the northwestern part of Victoria and extending into South Australia under the curve of the Murray River. Today it is wheatland and a source of considerable export income. The Mallee is a dry land. Every year in April farmers wait anxiously for the first of the winter rains. When the rains come, they hurry to plow and seed. Then, if the rain gods are kind and the Mallee's average annual allotment of rain actually falls at intervals through the winter, the harvest will be good. In some years, of course, the rainfall is far below average -drought and flood are themes of the Australian climate- and then the harvest is lean.

Wheat is a newcomer to the Mallee. The plant itself is a high-tech descendant of wild grains native to the Middle East. Wheat came to Australia with European settlers, who found the continent already inhabited by a hunting and gathering people. The ancestral aborigines probably arrived some 40,000 years ago, spreading at first along the coast of the continent, slowly learning to use the assortment of plants and animals that their new home had to offer. In time, small groups moved inland, some into the Mallee. Although poor in material possessions, these nomadic hunters and gatherers did not, as was once thought, live on the edge of starvation. They ate well, moving with the seasons and the chance falls of rain, exploiting the abundance this dray land had to offer. Widely dispersed in small groups, they thrived in places where European explorers died of starvation and thirst. The aborigines used the land in one way ; Europeans would learn to use the Mallee in other ways.

Australia's early European settlers did grow wheat. Not immediately, because the rhythms of the climate had to be mastered ; however, within a few years, there were harvests that gave hope that an European life could be lived in Australia. Yet, when the Mallee was entered in the 1840s and 50s, the newcomers were drovers bringing sheep and cattle rather than wheat.

The deterent was not the ability to grow wheat -the land and climate have long been suitable- but the problem of shipping it profitably to market. In the era before railroads changed everything, Australia's shortage of rivers made Mallee wheat impossible, for even the Murray has but few navigable tributaries and itself has no useful outlet to the sea. Inland products either had to walk to market or be carried by animal-drawn cart. Sheep and cattle opened up the interior because meat animals could walk to the slaughterhouses, and wool was valuable enough to bear the cost of overland transport. Wheat, which had only about one twentieth the value of wool per ton, could only be grown profitably within 50 miles or so of market, which in the Australian case meant the coastal cities and ports.

Making the Mallee a productive wheat-growing region required a long string of circumstances that included the domestication of wheat, the development of oceanic transport, the settlement of Australia by a wheat-producting people, and the development of cheap overland transport beginning with railroads. In short : a product, producers, markets, and access to markets. Wheat farming brought settlement to the Mallee. Where once aborigines and drovers lived their own sorts of lives, farm families now earn a living from wheat export.

We have options in space ; a choice of economic visions. The very high cost of transport means that space development -let alone settlement- is going to be a difficult and costly task. To make any of it happen in the near future, we must devote significant effort to the basic task of making transport to low-earth-orbit (LEO) much cheaper than it is now. But even if we are successful in that effort, and reduce the cost of launching a kilogram to LEO by a factor of ten or twenty, the expense would still be great enough to ensure that lunar products will be unable to compete in terrestrial markets. The Moon will have access only to customers in space, a limited market indeed.

We could choose a modest goal -a lunar science base and oxygen production facility, perhaps a martian outpost. With sufficient automation we could accomplish a great deal without the expense of establishing and supporting substantial communities. This is an Antarctic model with some resource development thrown in -an option requiring only a limited human presence. If this is the choice we make, I believe that significant extra-terrestrial communities would eventually arise, but only slowly.

Or we could make the deliberate choice to encourage settlement.

The high cost of transport demands that a lunar or martian community produces much of its own food, water, building materials, propellants, and so on. These basic self-sufficiencies are critical to reducing the financial pain to levels bearable to terrestrial taxpayers and politicians. There will be financial pain ; the first extraterrestrial communities are likely to be financially dependent for a couple of decades or so. However, while we are learning survival skills, developing the exports essential to self-support, and waiting for markets for lunar products to emerge, we could use a lunar science base not only to support valuable basic and applied research, but also to support a small but growing permanent population in much the same way that the penal establishement supported the growth of a private sector during the early decades of Australian settlement.

Let me take a few moments to sketch the history of this very important case.

Throughout much of the 17th Century Great Britain had dealt with the increase in crime that accompanied urbanization by transporting convicts to the New World. Once convicted a prisoner was transferred to a contractor who then took him to the American colonies where his services were sold for the duration of his sentence, usually seven or fourteen years. The system worked well from London's point of view since it carried convicts out of sight across the ocean, the cost was low, and it was generally hoped that either the threat of hard labor on a tobacco plantation (where most of the convicts wound up) would deter other criminals or, at least, that the high cost of passage would prevent ex-convicts from returning to England.

When the Colonies revolted in the mid-1770s they refused to accept any more convict shipments. In hopes that the war would soon be won, the convicts were housed in de-commissioned ships anchored in the Thames -to so-called hulks. Because the courts were sentencing roughly a thousand people a year to transportation, the hulks quickly became overcrowded. When the war ended in 1783 and both the United States and Canada made it clear that neither would accept convicts, the Pitt government found itself in desperate need of finding a solution to the convict crisis.

In essence, Botany Bay was a choice of last resort. It was generally recognized that a penal colony in Australia would be very expensive to establish and operate. However, the Pitt government had no other choice. It was hoped that convict labor could be used to make the colony self-sufficient. It was a false hope. The convicts were poorly motivated and generally unskilled in agricultural trades or any others for that matter.

However, circumstances led to the creation of a private sector which helped to greatly decrease the cost of Australian operations. The government was no more eager to have convicts return to England from Australia than from America, and so instructions were issued to the governor to grant land to convicts on completion of their sentences. Provisions were also made to grant land to members of the civil and military establishments in hopes that they too would linger in the colony and provide a core of trained personel. Grantees were usually given tools, seed, and convict labor to help them get started. As is usually the case, both groups tended to work harder for themselves than they did in government employ. By 1792, only four years after the arrival of the First Fleet, the colony was producing most of its own grain, and the private sector was the major producer. Because buying grain from these people both reduced the cost of feeding convicts and provided a market for the private sector, thereby encouraging their efforts, it soon became a regular practise for the commissariat to purchase grain (and meat, as it became available) at prices fixed by the governor but generally consistent the price of production and an adequate profit. As an incentive to the private sector and as a means of reducing the number of convicts being fed by the government, a substantial fraction of the convict population was "assigned" to private employers who were free to use the convict's labor provided that the employer fed, clothed, and housed him and provided a small wage.

One of the reasons that the system worked as well as it did was because it functioned in a growing economy. Had the number of convicts sent each year remained roughly constant, the impact of this "artifical" export market provided by commissariat purchase would have decreased each year, forcing the economy to rely on real exports quickly. Because Australia was a long way from potential markets, and because the lack of rivers delayed internal development, it was in fact several decades before a substantial export (wool) was developed. Had the Australian economy been cut loose early, development might have come even more slowly. However, for a variety of reasons related both to the increasing pace of industrialization and urbanization in England, and to the protracted conflict with Napoleon, the number of convicts sent tended to increase with time. In fact, throughout the first few decades, the Australian population grew at an average pace of 9 percent per year, with less than a third coming from natural increase. Expansion of the private sector was driven by the increase in the convict population. During the last decade of the 18th Century and the first decade of the 19th, roughly half the population was being fed at any given time by the government. Although the percentage fell to roughly 30% during the next decade, earnings from wool, whaling, sealing, and a variety of other activities were beginning to have some effect on the economy. By 1830 pastoralists, many of whom had been able to build up their flocks with moneys earned through commissariat purchases, were on the verge of producing the majority of Australia's export earnings. By 1840 the wool industry was secure enough that Australia could cut itself entirely loose from the penal establishment. Transportation to New South Wales came to an end in 1842, due mostly to appeals from the colony to London.

While I would certainly not propose that convicts be used as lunar settlers (I wouldn't want to have to be arrested for not paying a Paris hotel bill to get to go myself), a lunar science base could serve the same function as the New South Wales commissariat and become a buyer of goods and services from permanent lunar residents. Scientists and engineers on short-term visits to a lunar science establishment will have to eat and will need support services. A permanent community could be established to perform these functions. In the long run it will reduce the cost of lunar operations and, perhaps more importantly, this is one way of ensuring the development of a private sector during the time that work on developing paying exports goes forward. As with numerous settlers of the past, the first permanent residents of space will almost certainly need our help in getting established. We will probably have to provide passage, working and living space, and access to tools, power, water, and oxygen.

Making the Moon and Mars productive is going to be expensive, but I think the payoff will make the investment worthwhile. We have the solar system to gain.

This is the economic side of space development. But what about the trees ? I keep raising the issue of trees and grass and then wander off into economics. So let me finish with the landscape.

When European settlers went to the new places, they worked very hard to reproduce what they knew. In a few places, like New Zealand's South Island, the climate cooperated and the re-creation could be faithful at a surprising level of detail. But in most places they had to compromise and adapt to local realities. Everywhere they went, they changed the land. While the Mallee does not look like the English countryside, it is certainly not the land that the aborigines knew. In towns and cities, most of the environment was man-made; urban gardens and parks are pampered replicas of nature, often filled with immigrant trees and flowers.

Lunar realities preclude a farming frontier. A lunar settlement would look more familiar to city people than to farmers, more familiar to a pioneer settler of Sydney or Philadelphia than to settlers of the Mallee or the Kansas prairie. Still, there will be oddities : plants that grow in strange shapes under the influence of lunar gravity. Gardeners will fight against such tendencies, I suspect, until a native-born generation gives up and accepts the reality. The pioneering experience repeats.

The physical setting will be new; but I think that pioneers of other times and places would recognize the mood, the spirit of the enterprise.

Acknowledgements

This work was supported by the Los Alamos National Laboratory and written during a sabbatical year spent at the Department of History of Melbourne University, Melbourne, Australia and at the Department of History of the University of Alaska-Anchorage. My thanks to Goeffrey Blainey, Greg Denning, Lloyd Robson and Di Olle for their hospitality and guidance during my Melbourne visit, and to Stephen Haycox and Mead Treadwell in Anchorage.

Suggested Readings

[1] G.J. Abbott and N.B. Nairn, Economic Growth of Australia 1788-1821, Melbourne University Press, Melbourne, 1969

[2] Geoffrey Bailney, The Tyranny of Distance, Sun Books, Melbourne, 1966

[3] Geoffrey Bailney, Triumph of the Nomads, Sun Books: Melbourne, 1975

[4] Geoffrey Bailney, A land Half Won, Sun Books, Melbourne, 1980

[5] Brian H. Fletcher, Landed Enterprise and Penal Societey, Sydney University Press, Sydney, 1976

[6] Douglas Pike, Australia : the Quiet Continent, Cambridge University Press, Cambridge, 1962

COSMIC ACTIVITY: INTENSIVE WAY OF DEVELOPMENT

Arkady D. Ursul
Vice-president, member of the Moldavian SSR
Academy of Sciences

This year we celebrate thirty years'anniversary of the cosmic era that dates back to launching the first man-made Earth satellite in the Soviet Union. Space exploration had a considerable impact on science and technology, as well as on many spheres of human activity, promoting their development. At the same time breaking into space and the growth of its development scale led to new ideas concerning philosophical comprehension of the human role in the world and elucidated new ways of the humanity's future, as well as its cosmic direction of development.

As it is known, philosophical exploration of space has begun long before the beginning of cosmonautics. It were the thinkers of the Antiquity who dreamed of the unity of man and space, and their concept named anthropocosmism had a profound impact upon the development of the social and philosophical views of K.E. Tsiolkovsky, the founder of theoretical cosmonautics. It should be particularly stressed that K.E. Tsiolkovsky first created a definite philosophical concept of space exploration. He thought of a possibility of man inhabiting an environment different from that of our planet, which he named a free space. The scientist thought not only of the possibility but also of the reasons of people mass exit into space namely escaping earth catastrophies and overpopulation, creating better conditions of existence, the absence of infections, better power resources provision at the expense of Solar radiation, etc. So it is clear that it were ecological reasons that played a considerable, perhaps decisive, role in K.E. Tsiolkovsky's invention of the rocket means of reaching outer space.

K.E. Tsiolkovsky was one of the other pioneers of rocket technology and cosmonautics, who paid greatest attention to philosophical problems of principle concerning man feeling at home in space. Among the issues of philosophical and sociological range that were raised and to some extent developed by K.E. Tsiolkovsky I shall mention the following ones : bringing into light and discussing the reasons and aims of space exploration, the necessity of a cosmic point of view on human activity and the role of civilizations as well as of intellectual beings as such in the Universe, demonstrating the necessity of the production organization and development beyond the Earth and the Solar system transformation, the inevitable cosmic settling of mankind in the future and its subsequent immortality, as well as a number of other ideas.

J. Schneider and M. Léger-Orine (eds.), Frontiers and Space Conquest, 141–153.
© 1988 by Kluwer Academic Publishers.

Tackling the problems of cosmonautics and space exploration Soviet philosophers proceed from K.E. Tsiolkovsky's principal ideas and develop a number of new ideas that are connected with the humanity's contemporary space practices.

Nowadays in the USSR there is formed a special trend is philosophical research that is devoted to space exploration problems. Every year in Kaluga special sessions are held in the framework of the yearly K.E. Tsiolkovsky Readings. I have no time to mention all the problems and results of the philosophical space exploration achieved in the course of the research, so I will dwell upon some of them, those closely related to the theme of the present colloquium. But first of all I will say a few words on the philosophers'role in solving the problem of space exploration.

I think philosophers should not confine themselves to discussing any specific projects, neither the details of the future space exploration; instead, they should be working out the principles of cosmic mentality, as well as defining the main tendencies, natural laws and perspectives of cosmic activities. One should not concentrate oneself on some specific recommendations for space explorers, but should devote one's attention to generalizing the past, present and future of the theory and practice of the space flights. That is, in our opinion, philosophers must first of all reveal essential features and fundamental tendencies in the interaction between society and space that now form in that process a single socio-natural system of development. Of course, there exist the other philosophical problems of space exploration that are also mentioned at our interdisciplinary colloquium, but having general ideas on the contemporary and future cosmic activity is one of the priority philosophical tasks in this sphere. Practical actions devoted to priority directions in space exploration, concentration of efforts and means for concrete space programmes and projects depend on correct understanding of what cosmic activity is.

Our understanding of the role of space exploration for the life of society has undergone certain changes in the course of three decades. From first delights, amazements and the constatation of changing man's role in the Universe we later proceeded to discuss the necessity of a wider space exploration. Discussions of the kind were reflected in the world press; I'd like to mention, for example, a debate between scientists and writers in "Le Monde" published in July 1969. A positive answer biased towards the necessity of a wider space exploration was obtained as a result of practical cosmonautics achievements. Nowadays discussions have taken on a more local character, -namely, on the expediency of realization of this or that cosmic project or programme.

However, recently there have started widespread discussions as to the programme of Strategic Defence Initiative (SDI) advanced by the USA. After Reykjavik it became clear that it was because of SDI that the USSR and the USA didn't sign any agreements paving the way to elimination of the nuclear war as a menace to mankind's existence by the beginning of the third millennium. Our nearest future and, moreover, the fate of human civilization, depend on solving the problem of preventing space militarization at present. Man's exit into outer space and the latter's development marked a new era in interrelations between society and nature that open up a truly boundless scope for scientific and technical as well as for social progress and its continuous development. It represents of the triumph of human thought, but at the same time in the process of SDI programme development space means may turn into a tool of destruction of mankind and of the whole of the life on the Earth. This is a tragic contradiction on the way of civilization development at the end of the second millinnium, A.D.

Consequently, the problem of a possible destruction of mankind, the problem that until recently philosophers discussed abstractly relating it to a possibility of the Sun becoming extinct or to some other exterior reasons, -nowadays has acquired more than just a theoretical meaning. The future alternative (extinction or immortality) is now a question of vital importance influencing the fate of any person. That is why it has now become the most vital question of world outlook.

Whether man and mankind will or will not exist in the mode and space era is a vital political theme having, in our view, a cardinal philosophical meaning. Relating to the foundations of philosophy this theme connects the latter with practice. This represents one of peculiarities of contemporary social development under which a number of philosophical problems that seemed particularly abstract not long ago nowadays press for not only theoretical but also urgent practical, economic, scientific and technical as well as for political solving. We realize that solving the problem of preventing space militarization depends first of all on the practical decisions, on the future international situation, on the direction of the military danger development. None the less, we think that philosophers may also make their own contribution to the formation and dissemination of the new political thinking.

In our opinion, our humanistic mission consists precisely in combining the wisdom of philosophical thinking with an adequate comprehension of the nuclear and cosmic era realities, as well as in the active propaganda of the new political philosophy. The ultimate principle of contemporary humanism truly consists in preventing a thermonuclear catastrophy and the "star wars", -that is in human-race survival.

I suppose overyone will agree with this is principle, because all our discussion as well as further activity will prove to be meaningless if we are not able to prevent the danger of a nuclear holocaust and the shift of the arms race into space. So, the cosmic mode of thinking that made its way throughout centuries nowadays should be connected to the new philosophical and political ideas, synthesizing the principles of rational and humanistic activity. Thus, the strengthening of the principle of the humanity survival is simultaneously the acceptance of rational and humanistic activity. Thus, the strengthening of the principle of the humanity survival is simultaneously the acceptance of the principle of peaceful space exploration and a humanistic cosmic thinking.

Stepping up the mainline of peaceful space conquest sets a number of methodological problems solving which will influence the choice of strategy for further cosmic activities. There is no doubt that the strategy must be effective, i.e., first of all cosmic programmes' expenditures should have a maximum possible social and economic effect. There is quite enough data published in the press proving that cosmic means have proved economically effective, and the question is first of all about the satellites of applied purpose, about using the cosmic means in economics.

Cosmonautics'national economic effectiveness is on the one hand linked to the fact that from the very beginning cosmic means, that are a technological progress innovation, concentrated all the latest achievements of scientific and technical revolution. On the other hand, cosmic means are valid outside the planet, in a different ecological environment, which, as K.E. Tsiolkovsky realized it, has certain advantage over the Earth conditions. Cosmonautics is just an integral combination of branches of science, technology and organization. By way of new cosmic means it strives for the advancement of new spaces and objects (compared to the traditional Earth ones). That is, from the very beginning cosmonautics was an innovation and intensification process of social activity.

We regard intensification process, or the intensive way of development, as a priority utilization of the qualitative factors and resources, the radical activity reorganization with the aim of raising its effectiveness as well as speeding up progressive development. It is introducing new qualitative factors in the social activity that lead to growing effectiveness, and the more is the number of the qualitative factors combined into a single rational system, the more the growth is. The intensive way of the social activity development leads to its speeding up and going into the qualitatively new and higer positions. The essence of speeding up in the process of intensive development consists in the new growth of quality, i.e., the very speeding up appears indissolubly connected to quality.

Space exploration is on the most represented as widening up the scope of the limits of human knowledge and activity. This is undoubtedly rather important because extending the limits, volume, the field of activity, entering a new dimension are the most obvious results of cosmonautics advancement. Simultaneously this process of extensive growth is immanently interrelated with the qualitative changes in human activity, and understanding this appears no less essential for philosophical evaluation of the space era that is patterned on the exposing of the most general and essential features in space conquest by man. Let us proceed to considering the basic and in our view the most essential traits of the humanity's cosmic activity as the advancement along intensive progress line [1].

Apart from some other reasons, the so called information factor played rather a considerable role in the fact that cosmonautics appeared a powerful means of intensification. In virtue of its nature, information being a transferable, objectivizable component of reflection, is more than other factors capable of raising effectiveness of any kind of social activity and is the single unlimited factor of the former's intensive development. Space industry appeared to be one of the most information-consuming branches, and its product also appeared extremely repleted with the most advanced computers and the other information equipment. But the thing is not only that the space vehicles are repleted with information means but also that the chief result of the whole previous stage of space exploration consists in obtaining scientific information about space.

If we analyze deeply enough the reasons of mankind's starting space conquest we shall discover the fact that obtaining all kinds of information is almost the most important reason [2]. K.E. Tsiolkovsky in his time supposed that the basic results of space exploration consisted in obtaining unlimited space and Solar energy resources at mankind's disposal. All this remains valid, but at the contemporary stage of cosmic activity obtaining new scientific information about space and the space flights becomes the matter of primary importance. Obtaining and utilizing it proves to be the very innovation process for all the spheres of social activity. And though, besides information, we obtain both substance and energy from space as well as explore its farther reaches, these are not altogether the main cosmic achievements. At the end, all the other space merits and properties that are closely related to information are used, or consumed, by us first of all owing to information about, and from, space.

Thanks to its unique properties differing from other substance properties, information is a powerful lever of social progress. It is not by accident that information science is in the vanguard of contemporary intensificational processes in all the spheres of the social activity, the cosmic one included.

Space exploration, from this point of view, emerges as one of the "mechanisms" of civilization progress intensification that allows us to raise information content at the expense of the environment, to withstand entropy by advancing the sphere of negative entropy processes beyond the scope of the planet. Information characteristics of civilization do not reveal any "limits of growth" anywhere, and those setting such limits for the other indices of society's production forces development are hardly able to point at and to ground the informational "civilization stagnation" that is stipulated by natural laws. Owing to space conquest the possibility of infinite information accumulation in the social sphere serves speeding up the social progress which is the ultimate mode of the anti-entropial processes'manifestations in the Universe.

Cosmonautics' "information accent" will long determine the specificity of the cosmic activity development. At the same time the latter's intensification will comprise the utilization of the other factors of space and a space flight, namely, weightlessness, vacuum, radiation, etc., in order not only to model them on the Earth's surface but also to use them directly in space. The above mentioned space properties as well as its other processes, forces and conditions, represent the new natural intensification factors. It's true that with their help it is possible to considerably raise the effectiveness of a number of production branches and even to create branches of industry that are principally impossible on the planet. On the whole, following space exploration there will begin the process of its production development, which will entail moving more and more people outside the planet.

In the future, when the costs of space transportation will be reduced, it is supposed to shift there the base of not only ecologically the most unhealthy production branches, but also many other ones, first of all the power industry and other branches of industry whose space conditioned development will appear economically more effective because the conditions and forces of space will appear more natural for their large-scale intensified development at a time when there are a number of limitations on the planet related to negative reactions of the environment. While speaking of the process of the public production moving outside the planet we believe, that the primacy in the field will rest with the industrial production while the agricultural production turning to intensification adaptive strategy [3] will be longer tied to the Earth, because it is more natural here. Really, Earth plants consume the Solar energy, and the whole of the agricultural production, unlike industry, is based on biological processes and is "plunged" into biosphere. Outside the planet the existence of the agricultural production elements is principally possible, and its development in space was contemplated by K.E. Tsiolkovsky (nowadays there are biological experiments in this field carried out on board the orbital stations).

However, space conditions and those of a space flight (particularly weightlessness) create far more important obstacles (in spite of individual achievements, it is still not clear whether higher cultural plants can grow in conditions of weightlessness) for plant cultivation (to say nothing of cattle breeding). But while in the future more and more people will be going out into space accompanied by the industrial production development, agricultural production will also become biased towards space. But in our view it will comprise a qualitatively new production process compared to that of the Earth, and preliminary expertize attests to this fact [4].

That's why we suppose that at a certain stage of space exploration there will occur a "bifurcation" of material production into the terrestrial one that will be primarily agricultural, and the cosmic one (mainly industrial) between which there will be carried out an interchange of activities'products. Such a perspective of production development at large and that of agricultural production in particular, will appear the most natural and economically and ecologically expedient, because it is in the Earth-space bifurcation that the new possibilities of further advancement along the path of the all-round intensification and raising production effectiveness as a whole will be revealed.

Principally, we do not intend to "tie" the agricultural production development to the Earth. Later there will appear an agricultural space branch but, in our mind, it will be forming much slower than the space industrialization. Besides, cosmic intensification factors and the raising agricultural production effectiveness will influence the branch also in a terrestrial (but cosmicized) variant: one should attribute here remote probing, applying technological as well as medical and biological cosmonautics'achievements, and the systems of subsistence, particularly the experience of creating such systems on the basis of closed ecological cycles.

Our proposed hypothesis of terrestrial and cosmic public production "bifurcation" at a certain stage of their intensification still does't mean that the whole of industry will disappear from the face of the Earth. It is expedient just on the planet to retain and further develop the branches of industry that comprise the agro-industrial complex. The future production forces'growth will become both economically and ecologically expedient in space only where from the very beginning it will be possible to develop what has been termed an ecological production. The latter is perceived to be a certain artificial transformation of "*existing natural surroundings, sparing them with earlier impossible states and properties, producing new media that differ from the natural ones, in order to provide for smooth social progress*" [5].

The biosphere existence considerably reduces the possibilities of the ecological production on the planet, though does not exclude it completely. At the same time, space having no other biological and social spheres, there appears a possibility of creating new media on a large scale that are suitable for alike but differing from the natural ones, right up to forming artificial space human dwellings on the basis of the Jupiter planet group. From the very beginning the space ecological industry organization may acquire a planned character and become a general trend in eliminating ecological dangers and disorders. Further sociosphere growth, and also the social progress, will take place already not as a result of the Earth biological and geographical spheres'degradation but at the expense of less organized space inanimate nature that will naturally comprise the wastes of the material production (first of all power production) which will become a public cosmic process. Space ecological production will not represent a separate kind of the material production but will become its essential point organically tied to its other kinds and the branches of human activity that provide for its maximum socio-economic effectiveness.

On the whole, ecological production growth is not suitable everywhere but only in the ecological conditions where there is a great difference in development stage between society and nature. It is such a difference that is present in space conditions under which human existence is in principle impossible with no ecological industry (i.e. exo-industry) developed. That is why the intensive way of the cosmic activity development will be determined to a considerable degree by the necessity of creating human life conditions in space, that is, by ecological industry requirements.

While speaking of the natural factors in the cosmic activity intensification, we mean first of all (and this is prime importance nowadays) the forces, processes and conditions of the space inanimate nature. We think that in the future a change may occur in intensification factors situation. Truly, outside the Earth there is a possibility presumed of a different life and cosmic civilizations having fundamental development laws in common with terrestrial life and the humanity. Provided they really exist and we are able to establish information (if only) contacts with them, this would result in a great impulse in the process of production intensification at the expense of extraterrestrial biological and social intensification factors.

We doubt establishing information contacts with extraterrestrial civilizations before the end of the current hundred years. But it is necessary to develop this research trend in theoretical and practical aspects, because it will contribute to a more profound understanding of mankind as a space civilization as well as an essential unity of mankind in the face of space and a different mind compared to that of ours [6].

It is possible that bringing the arms into space by earth men may be regarded as a threat by representatives of an advanced space civilization if they appear in the vicinity of our planet. In case of the arms race transfer into space the fate of out civilization will appear not only in the hands of the humanity itself. There are many reasons according to which we do not detect extraterrestrial civilizations and thus do not co-operate with the "social space". To these we should add that cosmic civilizations'transition to intensive development presupposes various rational self-limitations in the name of the social progress above all. If other space civilizations could evade a threat of space cataclysms and self-destruction, it is doubtful they would spend their power resources on sending a non-directed signal carrying information of their existence. Besides a colossal power expenditure, such a signal is fraught with negative ecological consequences. It is possible that other space civilizations also set the limits to the ecological niche of their spreading throughout space. They should undoubtedly give primary consideration to their internal development processes setting about interplanetary communications only when it proves also economically effective.

Going out into space in the face of its representatives our terrestrial civilization also pays great attention to solving its own global humanity problems, which means that the means of cosmonautics are aimed at solving the problems of ecology, food, power, health care, education, etc. There is no global problem, nor any economic, scientific and technical nor any other contemporary humankind problems, where the cosmic means did not exercise a positive influence, naturally, in their peaceful utilization. At the same time, means of cosmonautics aggravate some global contradictions, first of all the problems of war and peace, as well as the ecological problem, because they are used with militaristic purpuses.

Owing to cosmonautic's peaceful advancement, there takes place a more profund cognition and the rational development of the terrestrial nature, "extraction" and "objectivization" of scientific information about space, drawing into production of the conditions and processes of the near space for the welfare of mankind. This in itself is conductive to forming more close interrelations between society, the planet and space, and the socio-natural system "mankind-the Earth-the Universe" is formed, in which many global problems may be solved that are raised by scientific and technical revolution and by the social progress gaining speed. At the present stage of space exploration cosmic means application for accelerating peaceful scientific and technical as well as the socio-economic progress on the planet appears the top priority task. It is this concept demonstrating the necessity of placing man and mankind as well as the latter's peaceful and progressive development in the focus of the aggregate cosmic activity, that was termed "anthropogeocosmism", or "sociogeocosmism" [7].

From this concept it appears that for a long time the centre of the newly forming ecosystem comprised of the Earth and continually growing realms of the nearby space will rest with our planet, by force of the fact that it is "inhabited" by humankind. Further on, however, in a comparatively remote, in our mind, historical perspective, there will appear some other cosmic activity centres equal with the Earth: both artificial structures, as suggested by G.O'Neil, and human settlements on the natural celestial bodies suitable for the purpose (the Moon, Mars, satellites of the Jupiter group planets). But in the presence of such new centres the Earth will for a long time remain a dominating centre guiding human civilization's space activities.

These problems were previously tackled by us in a number of works [8], and we also suggested the so-called matrices of the human activity cosmicization that give an idea of imparting it a cosmic content, as well as of the trends in cosmicizing its basic components. One cannot imagine the emergence of the new research and activity centres as a decentralization process only, as the process of activity centres vanishing. Instead, it is a relative overcoming geocentrism in a sense of widening up the scope of exploration followed by practical activity. Thus, such overcoming does not yet mean that the Earth ceased functioning as a centre of the research and theoretical activity, on the contrary, breaking into space convinced us of the unique character of mankind activity and inhabitancy, in the necessity of preserving it for the future generations.

As the humanity's cosmic activity will take on an intensive mode of development, as it was mentioned above, it will not at first provide for a radical transformation of the space environment (e.g. dismembering the Jupiter group planets). It is important not only to transform space but also to plan the activity so that to reduce the transformation of the nature to a minimum, or even to improve it, in order to meet the needs of the social progress as much as possible. Here the principle of the socio-economic effectiveness is combined with that of the ecological expediency. We suppose that in the process of a subsequent large-scale space transformation (and such an activity was termed cosmocreatics) it is advisable to transform the natural space processes only in case of emergency, trying to use them in intensification processes, so to say, in their natural aspect. This principle of minimizing the necessity of the ecological environment transformation that must be fully developed in space, is also quite valid for further activity on the Earth, provided we want to avoid the threat of a global ecological catastrophe and to preserve the biological and social spheres.

None the less, even applying the principle of ecological and cosmic activities, it is possible that in the remote future the humanity together with similar space civilizations will become a powerful factor of the Universe evolution, if one supposes an indefinite duration of the cosmic civilization existence. In this connection, there appears a question as to the limits of space exploration in an astronomical perspective. Taking into account the possibilities of accelerating scientific and technical progress it is impossible to set beforehand any frontiers for the advancement in the Universe. However, it is clear that any cosmic civilization process and the whole of their possibly existing system should always remain just a certain element of the Universe that is represented as a meta-ecological environment. A civilization in space must be enveloped in the biosphere taken out there the volume of which will exceed that of the civilization organic body. This follows from the fact that there must be preserved a correlation between the social and biological spheres that is characteristic of a producer and consumer correlation. That is, there exists the so-called "*negative entropy pyramid*" [9] which is valid also for the correlation of the lower levels of substance development, namely, chemical and physical objects. Such correlations reveal themselves in the process of the study of dissipative structures carried out in the framework of a new research trend termed synergetics. Applying it to a cosmological scale we shall be able to draw nearer to the answer to the question of possible limitations of our expansion in the Universe. It is evident that we are laying down a research programme in the framework of the space synergetics that has not yet come into existence, than answer the question of the frontiers of space conquest. More than once we recurred to this problem [10] and in the course of philosophical analysis we did not reveal such insurmontable obstacles. Neither did we reveal any time limits to human existence, excluding, however, the possibility of mankind's thermonuclear self-destruction. As K.E. Tsiolkovsky believed (and we additionally substantiated this concept in such works as "Space exploration" and "Humanity, the Earth, the Universe"), owing to space exploration there opens up a perspective of unlimited social progress, or human civilization immortality.

Speaking of the perspectives of space exploration in terms of advancing cosmotechnical, or astroengineering activities one should pay attention first of all to ecological situation and the conditions that set the limits not so much to the cosmic activity as to the extensive expansion. Ecological norms and limits undoubtedly exist not only on the Earth but also in space where they differ from those on the planet. These limitations are indicative of the transition to the cosmic civilizations'intensive development. Cosmocreatics in an intensive cosmotechnical activity on an astronomical scale. But the question is not so much about the radical and global change in the structure of the Universe observed today, as about cosmo-ecological activity under which the intensive civilization progress will be accompanied by the minimum ecological environment destruction.

All these problems will be tackled at a junction of philosophy of space and social ecology also turning to space. Recently one comes across the concept of space ecology more and more often. There is yet no clear notion of the subject of this new research discipline. Some scholars suppose the discipline should study "The conditions of the prolonged or constant existence of man as well as of the animal world and the vegetable kingdom organized by him beyond the Earth limits" [11]. However, it would be quite insufficient to reduce to this task the ecological research problems in connection with the advancement of the space branches of science. Cosmic means application for solving ecological problems on the Earth is also the task of the cosmicizing ecological disciplines.

In our view, the subject of ecology expanded at the expense of the cosmic objects should comprise also an ecological cosmology aspect, as well as a clear notion of the properties and conditions of space as an environment necessary for the emergence of life and society, and not only the local conditions but also the global ones, i.e., in the spirit of the anthropocosmological principle that is being advanced nowadays [12]. It is possible that such a task, as well as the study of the ecological aspects of the extraterrestrial civilizations problem, should be attributed to the sphere of astro-ecology and we term the latter by analogy with astrosociology which studies the social aspects of the extraterrestrial civilizations problem. Here we make distinctions between space ecology and astro-ecology along the lines of the former being related to space exploration and cosmonautics and the latter being related to the complex of the astronomical knowledge. In any case, whatever are the terms the idea of the ecological problems existing in space sciences seems quite correct. And this is the more argumented from the point of view of introducing the ecological approach as a general scientific one being simultaneously a fragment of general scientific knowledge [13] that penetrates all the areas of knowledge including the disciplines related to the inanimate nature (synergetics existing nowadays, in particular, is undoubtedly directly related to the "inanimate nature ecology"). So, the expansion of the subject area of the ecological knowledge comprises two reverse processes, these of ecology cosmicization and the space sciences ecologization. These processes reflect dialectically interrelated processes of the social activity: those of the Earth ecological activity cosmicization and the space activity ecologization. In our mind, both processes appear to be the aspects of a more general tendency manifested in the social activity advancement along the lines of intensive development while being its characteristic indication.

The transition to a primarily intensive mode of development and a gradual aspiration for a more comprehensive and total intensification is, in our mind, a fundamental regularity of the social progress that is independent of any concrete forms of manifestation of sociality in various space areas. Its fundamental and invariant character is related to the fact of the absence of any other mode of rational development except of course the extensive one that is less effective and is inherent in civilization process only at the earlier stages of their technological activity and that also inevitably adds to some extent to the trends of the intensive development. That is why we suppose that not only the advanced states on the planet but also highly developed space civilizations rationally interacting with their environment are the intensive-type civilizations widely using qualitative factors and resources for their technological progress and providing for a maximum effectiveness in social and economic terms, as well as for nature preservation and ecological effectiveness. Cosmic activity is the intensification process that in the course of its humanistic advancement considerably contributes to development acceleration and a sure achievement of the highest qualitative progress gains.

References

[1] A.D. Ursool, A.I. Dronov, Kosmonavtika i sotsial'naya deyatel'nost', Kishinev, 1985

[2] A.D.Ursul, Osvoyeniye kosmosa (filosofsko-metodologicheskiye i sotsiologicheskiye problemy), Moscow, 1967

[3] A.A.Zhuchenko, A.D.Ursul, Strategiya adaptivnoi intensifikatsii sel'skokhoziaistvennogo proizvodstva,-Kishinev, 1983

[4] A.V. Rusakov, G.G. Rusakova, Rol'kosmicheskikh issledovanii dlia zemledeliya budshchego, Ekologiya i zemledeliye, Moscow, 1980, p. 226

[5] Ye. T. Faddeyevb, Problema ekologicheskogo proizvodstva, Puschino, 1980, p. 13

[6] V.V.Rubtsov, A.D.Ursul, Problema vnezemnykh tsivilizatsii, Filosofsko-metodologicheskiye aspekty, Kishinev, 1984

[7] A.D. Ursul, Chelovechestvo, Zemlya, Vselennaya, Filosofskiye problemy kosmonavtiki, Moscow, 1977

[8] V.I. Sevast'yanov, A.D. Ursul, Era kosmosa: obshchestvo i priorda, Moscow, 1972, pp. 51-60
A.D. Ursul, Chelovechestvo, Zemlya, Vselennaya, Filosofskiye problemy kosmonavtiki, Moscow, 1977, pp. 73-80

[9] K.K. Rebane, Energiya, entropiya, okruzhayushchaya sreda, Tallin, 1984, pp. 32-35

[10] A.D.Ursul, Nekotorye filosofskiye voprosy osvoyeniya kosmosa, Moscow, 1964 ; idem. Osvoyeniye kosmosa, pp. 189-200
[11] Yu. A. Shkolenko, Ekologicheskiye aspekty obzhivaniya kosmosa, Filosofskiye problemy global'noi ecologii, Moscow, 1983, p. 330
[12] A.D. Ursul, R.A. Ursul, Evolutsiya, Kosmos, Chelovek (obshchiye zakony razvitiya i kontseptsiya antropocosmizma), Kishinev, 1986, pp. 128-154
[13] A.D. Ursul, Filosofiya i integrativnoobshchenauchnye protsessy, Moscow, 1981

WILL SPACE CHANGE HUMANITY?

Ben Finney
Department of Anthropology
University of Hawaii
2424 Maile Way, Honolulu, Hawaii 96822 USA

Introduction

In 1956 the philosopher Hannah Arendt delivered a series of lectures at the University of Chicago in which she considered the human condition from the vantage point of what were then "*our newest experiences and our most recent fears*". Three great events, she said, had shaped the modern age and determined its character: first, the discovery of America and the ensuing exploration of the world; second, the Reformation and the social and economic transformations that followed ; third, the invention of the telescope and the revolutionary perspective on the earth and the heavens promoted by astronomy. Before, however, these lectures could be published another great event occurred: on October 4, 1957 the first Sputnik was launched, inaugurating the space age.

When in 1958 Arendt published her lectures in a book entitled The Human Condition on the very first page she called the launching of sputnik an "céent, second to no other" [1]. She was disturbed, however, that this event was hailed by some as the first "*step toward escape from man's imprisonment on the earth*" for to Arendt "*The earth is the very quintessence of the human condition.* "*The most radical change in the human condition we can imagine*" Arendt declared, "*would be an emigration of men from earth to some other planet*". But, in her book, she did not develop this theme. Indeed, at that time when the space age was just dawning it would have been virtually impossible to consider seriously how developing activities in space might affect the human condition. Now, however, almost thirty years after the first sputnik, we are in a better position to consider the question posed by the title of this paper, 'Will space change humanity?' and to develop Arendt's prescient forecast that migration from Earth would radically alter the human condition.

Perhaps the most thorough post-sputnik exploration of the impact of space activities on the human condition to appear so far in English is the book The Universe and Civilization written in the late 1970s by cosmonaut Vitali Sevastyanov, and philosopher Arkady Ursul and sociologist Uyri Shkolenko. Quoting liberally from Marx and Engles as well as from modern space scientists and commentators from East and West, these Soviet authors examine the impact of space developments on humanity. Their point of view is social and philosophical, and they

J. Schneider and M. Léger-Orine (eds.), Frontiers and Space Conquest, 155–172.
© 1988 by Kluwer Academic Publishers.

label the changes in society and human consciousness that follow from our entry into the space age as manifestions of a process they call cosmicisation. Concrete examples of cosmicisation which they discuss include the development of new vistas in science opened by the use of space probes and satellites for basic and applied research, the flow of benefits to mankind from the employment of satellites for remote sensing, communication, and navigation, the promotion of an ecological awareness of the fragility of our biosphere inspired by those first views from space of our lonely blue planet, and the evolution of a new social framework and production relations as we begin to exploit the resources of space.

In addition to analysing this process of cosmicisation, Sevastyanov and his colleagues also discuss the colonization of space. They argue that the settlement of the solar system, and then interstellar migration, is inevitable, and they castigate such Western writers as C.S. Lewis and Loren Eiseley for their insistence that man should stay on Earth. Yet, they do not explore the full implications of the spread of humanity into space. While they stress (as does Arendt) that it was the Russian space pioneer Konstantin Tsiolkowski who proclaimed that it is man's destiny to spread into space, these Soviet analysts declare that this original conception has now "*been supplemented by a kind of 'feedback' aimed not at escape from Earth but at a closer man-made bond between Earth and outer space*" [2]. In fact, Sevastyanov and his colleagues devote most of their attention to this Earth-space bond. Indeed, their largely geocentric approach is inherent in their definition of cosmicisation as "*the effect of space factors, forces and processes on various domains of the conscious and goal-directed activity of men on earth*".

There is a need for a perspective that transcends the Earth-space relationship and considers the possibilities inherent in the spread of humanity through space. The discipline of anthropology, embracing both the biological and cultural evolution of mankind, provides such a perspective. In this paper I look at the impact on humanity of the impending migration into space from the perspective of an anthropologist concerned with the human future. As a philosopher, Arendt referred back to the thinkers of classical Greece and their world view in order to develop her analysis of the changing human condition. As an anthropologist, I reach back some five million years to the dawn of humanity, survey the events in human evolution -both biological and cultural- that have brought us to the threshold of space, and then attempt to project the trend of human evolution far into a future in which our descendants will have learned to live and prosper in space.

Let me introduce my analysis by stating the thesis of this paper. Man evolved as a exploratory, migratory animal. The human lineage began in East Africa, slowly expanded over the African continent, and then into Asia and Europe, and finally to the Americas, Australia and the islands of the sea. Our ancestors were able to spread from their tropical homeland through developing technology to travel to and survive in a multitude of environments for which they were not biologically adapted. Migrating into space, and the development of transport, life support and other systems which will be required for the maintenance and spread of human life there, represents a continuation of our terrestrial behavior, not a radical departure from it.

Nonetheless, although colonizing space may be a very human activity, such a move would radically transform humanity. By following our exploratory bent and by exploiting our knack for innovation, we will be entering into an era of accelerated evolution, first cultural and then biological. Faced with new challenges, forced to adapt to myriads of new environments, and dispersed in discrete communities scattered far and wide throughout space, humanity must change, again and again. If we migrate into space the human condition will be utterly and inalterably transformed.

To develop this thesis I will first outline the key elements in human evolution and the spread of our species over the globe, then examine the motivation behind our expansion, and finally attempt to look into the future to see the likely consequences of migration into space.

Human Evolution and Global Expansion (*)

"*The first giant leap for mankind*, to borrow a phrase from Neil Armstrong, *was the descent from the sheltering trees of the tropical forest to the grasslands of the savanna made by our distant ancestors who in so doing virtually set human evolution in motion*" These were literally the first steps toward humanity, for they were made on two legs instead of four. This revolution in posture left the forelimbs free to carry objects and above all to make an employ tools, for it is this growing technological capacity that has made our evolution so unique.

We humans are hominids, the surviving species of a lineage of erect-walking primates. The available evidence indicates that the hominid lineage evolved in Africa, specifically in East Africa. As Darwin [3] pointed out over a century ago, our closest relatives, the Chimpanzee and the Gorilla, come from Africa, and it is in East Africa that the oldest bipedal fossils have been found [4]. These date back some three to four million years ago, but are not thought to represent the oldest of our hominid ancestors. New techniques of comparing the chromosomes, serum proteins and hemoglobins of man and the apes, and of calculating the evolutionary distance between them, indicate that the appearance of the first hominids took place some five million or

---

(*) This section is adapted from

Finney, B.R. and E.M. Jones, 1983
From Africa to the Stars: The Evolution of the Exploring Animal.
In Space Maunufacturing 1983, J.D. Burke and A.S. Whitt (Eds.), 53, Advances in Astronautical Sciences, American Astronautical Society, San Diego, 85-103

so years ago [5, 6 & 7].

If the first known fossil homonids, those of the genus Australopithecus are any indication, our first bipedal ancestors were not mighty hunters who slaughtered game with spear and club. Although her male counterparts may have been somewhat more robust, the famous Lucy, the earliest preserved Australopithecus skeleton, stood barely over a meter in height, probably weighed under 30 kilos, and lacked the powerful canine teeth and other specializations of a predator [8]. In all likelihood these early hominids depended primarily on nuts, berries, tubers and other wild vegetable foods for their sustenance, supplemented by eggs, insects and perhaps some small animals. Although definite archaeological evidence is lacking, it is thought that these small hominids used rudimentary tools -wooden digging sticks, stone pounders, simple containers of bark and skin, and the like- to help them gather, carry and store their food, and that such modest tools marked the beginning of technological development [9].

Yet, despite the technological beginning, and the postural revolution that made it possible, Australopithecus apparently did not expand beyond the savannas of Africa. Evidence for migration out of Africa only comes with the evolution of a new genus and further technological advance.

While Australopithecus may have stood erect, its brain remained small, averaging around 500 cc., hardly bigger than that of contemporary chimpanzees [10]. Then, starting about two million years ago, the evolution of the brain began to accelerate. The first evidence comes from a skull of around 650 cc. found by the late Louis Leakey at Oldavai Gorge, Kenya [11]. Both because of the larger brain, and the fact that for the first time stone tools had been found in association with fossil remains, Leakey classified this creature as the first known representative of our genus, Homo, and gave it the species name of habilis to denote in undoubted handiness with tools. Although recent discoveries indicate that Australopithecus used rudimentary stone tools [12], by the time of homo habilis the distinctly human synergy between the development of increasingly sophisticated tools and the acceleration of brain development was apparently well underway [13].

Yet, these early representatives of our genus were not the ones to migrate from Africa. The first human fossils found outside Africa belong to representatives of a new species, Homo erectus. This new species had a larger brain; its cranial capacity varied from 775 cc. to 1225 cc, or into the low end of the range for contemporary Homo sapiens [10]. Homo erectus employed a more highly developed stone technology, apparently inventing, for example, the art of chipping stone tools on both sides to make a sharper edge. With better weapons, Homo erectus began to hunt more, shifting the emphasis from gathering wild vegetable food to preying on large animals. This shift also apparently involved more sophisticated forms of social organization. For example, as can be reconstructed from excavated hunting sites, these hunters employed guile and teamwork to drive animals as large as elephants into bogs where they could be slaughtered with spear and club, and then butchered with finely chipped stone cutting tools. With this new hunting way of life, and the technological and social advancements that made it possible, Homo erectus was able to migrate out of Africa and across the breadth of Eurasia, where fossil remains have been found spread from Western Europe to China and Indonesia.

In Europe and Asia this tropically-adapted animal must have found it difficult to survive the cold, particularly during times of glacial advance. Yet, the archaeological evidence indicates that bands of Homo erectus roamed far to the north to exploit the game-rich grasslands there [10]. Cultural rather than biological evolution was the primary mechanism for survival. Archaelogical sites indicate that they began to build rudimentary shelters and wear animal skins for protection against the cold. More significantly, fire-blackened hearths indicate that thee hunters had learned to control fire -one of the most important innovations in cultural evolution.

Yet, for all his hunting skill and cultural ingenuity during the million or so years this species flourished, Homo erectus does not appear to have spread beyond the linked continents of Africa, Asia and Europe. The archaeological evidence indicates that the move to the Americas and Australia required further cultural development and the appearance of a new species, Homo sapiens.

Exactly where and when our species, originated has not yet been definitively established. Recent research on mitochondrial DNA indicates that our species may have arisen in Africa some 2000,000 years ago, and then spread over that continent and Eurasia, replacing erectus populations [14]. Whatever the exact origin of Homo sapiens, and the position of Neanderthal Man in this evolution, the important point in this narrative is that Homo sapiens were the first to populate the hitherto empty continents.

The lowering of sea levels by 80 to 100 meters during the last glaciation provided the opportunity for human expansion into the new continents first, by exposing the continental shelves so that Siberia and Alaska were joined by a land bridge, and second, by making Indonesia into an extension of Asia reaching out almost to the shores of a great continent composed of Australia, New Guinea, and surrounding continental shelves. All that had to be done was to walk across the land bridge into the Americas, and to float across the short stretches of ocean separating Asia from a glacially enlarged Australia [15 & 16]. Yet, previous glaciation had similarly lowered sea levels without any migrations taking place. The crucial ingredient was the evolution of human cultural capacities and techniques [17]. Refined hunting tools, tailored skin clothing and other survival gear enabled hunters to penetrate far into the Arctic, and then to follow prey across the land bridge that then linked Asia and North America. Similarly, once simple rafts and rudimentary techniques for living off the sea and coastal resources had been developed, people could cross over to Australia. Although archaelogists are divided on exactly when these movements took place, most would probably agree that they probably took place between 50,000 and 20,000 years ago.

Whatever the exact dates of the origin of our species, and its spread over the continents of the globe, the important point is that in surmounting tropical and arctic barriers, and then in spreading over the forest, the mountains, the plains, the deserts and the jungles of the world, these accomplished wanderers utilized the unique human ability to adapt culturally to new environments. Building on the biological foundation of erect posture and brain expansion, our more recent ancestors developed further the human capacity to invent and apply technology to make human existence possible from Africa to the Americas, and from the tropics to the arctic. Where other animals have to adapt biologically to move into environments radically different from the ones in which they evolved, Homo sapiens, the hairless biped from the African savanna, could adapt culturally.

To claim, however, that late Pleistocene Homo sapiens spread over the entire world is to ignore the fact that we live on the water planet. Seventy percent of the earth's surface is water, and it is only in comparatively recent times that man learned how to sail well enough to colonize oceanic islands and use the sea for migration and trade. Further developments, both in terms of the technology of seafaring, and new forms of social organization needed for undertaking overseas expansion, were required to complete the settlement of our globe.

The first people to sail far out to sea were the Polynesians whose expansion begins along the shores of South China and Southeast Asia some 5,000 years ago and with the discovery and colonization of every inhabitable island in a vast oceanic realm the size of most of Europe and Asia combined [18]. This Polynesian odyssey foreshadows the coming expansion into the archipelagoes of space. Just as we are striving to develop means to spread human communities throughout space, so did the Polynesians, through developing large sailing canoes, precise methods of navigation, a portable system of agriculture which could be implanted on the virgin islands they found, and a social organization adapted for oceanic colonizing, spread humanity far and wide through a then alien environment [19]. Yet, however superbly adapted the Polynesian system was for colonizing uninhabited islands, once they had settled all the available islands in the Pacific, their expansion stopped.

The true discoverers of the global sea were those European navigators who, in learning how to sail between continents and eventually around the world, were the first to realize that there is but one ocean and that it could be used as a highway to connect hitherto isolated lands and populations [20]. The story of the development of ships and navigation, and the military, political and commercial instruments for expansion, is well known. Here I wish to emphize the global consequences of this discovery, and to stress that these amounted to much more than the territorial and commercial enlargement of a handful of European powers. This maritime age of discovery led directly to the bringing together of scattered branches of humanity into one world economic system, in effect completing the first phase in human expansion and setting the stage for the second.

The Exploring Animal

What can be said about this global expansion of humanity beyond the apparent truism that life expands to fill all niches, and the corollary than through both biological and cultural evolution humans are uniquely adapted for expansion? Is there something more to the spread of humanity than just the capacity to develop technologies and forms of social organization for expansion? I raise these questions in order to focus on the premise that we are by nature an exploratory, migratory species, and to examine the issue of motivation.

Let me begin to answer these questions by posing another one pertaining to the oceanic, phase of global expansion. Why did a few small countries on the western fringe of the Eurasian land mass initiate this reunification of humanity? Contemporary historians have updated the old explanation of rapacious Western greed by pointing out that the economic crisis affecting late feudal Europe drove its sailors out to sea, and forced its princes and bankers to support them [21 & 22]. Yet, the Portugese, Spanish and other European nations that turned to the sea were not unique in their economic problems. It is their creative solution, not their problems, which stands out. I side with earlier historians who stressed that the drive to explore the ocean and seek new routes and lands was crucial to the discovery of the sea and its consequences. Similarly, I am impressed with how the exploratory drive of the Polynesians led them to expand faster and farther into the Pacific than population pressure could ever have driven them.

While population growth and the search for resources have undoubtedly been crucial to human expansion, as these two maritime cases suggest, humans exhibit what the biologist Dobzhansky [23] has called an "*urge to explore*". At times in history, this urge has driven man farther and more widely than population crisis, economic exigency, or other situational determinants could have pushed us. Although this urge to explore is obviously not manifest in all individuals, nor in all cultures and ages, for any species to expand it only takes the adventuresome few to lead the way. That there have been sufficiently daring cultures, and individuals within them, to expand our species globally is evident. My bet is that there will be enough to spread mankind into space.

Exploration is not, of course, unique to humans. However, in most species the juveniles do the exploring, investigating their environment before settling down on a limited geographical range from which, as adults, they seldom stir [24]. Modern man follows a similar pattern of juvenile exploration before settling down to the routine of adult life, yet some adults do not give up their exploratory bent and, in fact, make a career of it. Columbus did this through force of personality ; by the late 18th Century maritime exploration had matured to the point where Captain Cook could claim to be "*employed as a discoverer*". Now, there are those who make their living by exploring the stars and planets through telescopes and robot spaceships, and a growing corps of astronauts, cosmonauts and spationauts are exploring space directly. We are an animal that has turned a juvenile characteristic into an adult profession.

This development may be as much a part of our genetic evolution as it is of our cultural evolution. Adult humans resemble juvenile apes in their large grains, globular heads and lack of protruding muzzle, a comparizon which has led to the theory that we have become large-brained humans through the process of neotony -through changes in growth rates that have acted to preserve foetal and juvenile characteristics far into maturity [25 & 26]. At birth infant apes and humans are much alike in brain size and facial configuration. But, while the sutures close early in an ape's skull, and its brains grows little as its brow thickens and it develops a protruding muzzle, our sutures remain open and our brain continues to grow through a prolonged infancy and adolescence. We then reach maturity with under-developed jaws and teeth tucked inconspicuously under the bulbous forehead of a child, a neotonous process that has in relatively short period nearly trebled the size of our brain.

Following the ethologist Konrad Lorenz [27] we may extend this theory of human evolution through neotony into the behavioral realm. Our hypertrophied urge to explore may be regarded as a behavioral manifestation of our neotonous development, for we -or at least some of us- retain the juvenile penchant for exploration and investigation of our environment far into adulthood. According to Lorenz, this retention has served us well, for from it comes our inquisitiveness into the nature of things as well as our constant search for what is over the horizon -in other words, it is the source for both science and exploration. Through this neotonous process, perhaps modulated by the mutation of only a few regulatory genes, we have become a most enquisitive, exploring animal.

The Colonization of the Solar System

We now stand on the threshold of space. Since the launching of the first unmanned satellite, men have walked on the Moon and spent over seven months in orbit around Earth, and plans are now being developed for the exploration of Mars, and the eventual establishment of permanent bases there and on the Moon. The dream of Konstantin Tsiolkowski, of Robert Goddard, and of Hermann Oberth of the establishment of colonies in space could be realized within the next century or so -unless that same rocket technology that makes space expansion possible brings nuclear destruction to the Earth.

Let us, however, be optimists and assume that a "*nuclear winter*" does not end human life on Earth [28], and that other disasters, manmade or natural, do not eliminate the human race. The real question then, is not whether attempts will be made to colonize space, but how farreaching and successful those attemps will be.

It may turn out that human outposts in space will be few in number and contain only small groups of temporary inhabitants. The harshness of airless, intensively radiated space environments, the difficulties in developing artificial ecological systems, the loneliness of life in confined habitats so far from friends and relatives and familiar sounds, sights and smells, and other unforeseen difficulties may inhibit the colonization process. In particular, it may not be possible to develop thriving communites of men, women and children which would form the nucleus of an expanding solar system population. The future in space might therefore be one in which the resources of the inner solar system are exploited by the Earth with minimal human presence in space. Solar power stations beaming energy to Earth, mining bases on the Moon, Mars and selected asteroids, and space stations assigned to scientific, industrial and military tasks could be part of such a future, for they could be operated with a combination of advanced robotic techniques and the services of crews which shuttle back and forth between space and Earth on temporary tours of duty. But there would be no real migration into space.

If that turns out to be the case, then the impact of space developments on humanity though considerable would nonetheless be limited. To be sure, the process of cosmicisation, in the sense used by the Soviet authors cited earlier, would proceed. Our understanding of the evolution of the solar system and therefore the Earth would improve immensely. The infusion of resources, which would include manufactured goods as well as power and perhaps some especially valuable raw materials, would materially enhance life on Earth, as would the reduction of pollution by the location of power generation and other industrial processes away from the surface of our planet. Similarly, more intensive use of space for remote sensing communication and navigation would bring other practical benefits. In addition, the enhancement of scientific understanding and the common use of space resources, material and informational, would undoubtedly affect our view of the Earth and our outlook on the human future. But, without actual settlement in space and the expansion of human communities there, the impact of the space age would stop short of being revolutionary. In this situation the geocentric focus of the concept of cosmicisation would be most appropriate.

Suppose, however, that the barriers to space colonization can be overcome. If ways can be found to avoid the damaging effects of cosmic radiation, if we can develop extensive space biospheres capable of permanently supporting great numbers of people, and if emigrants from Earth can learn to conquer feelings of fear, loneliness, and nostalgia, then the wholescale colonization of the solar system could follow. This colonization could proceed through the settlement of the Moon, Mars, the asteroids, and other natural bodies [29, 30, 31 & 32], or through the use of resources from these bodies to build, as Gerard O'Neill [33] has proposed, large artificial habitats in interplanetary space. Most likely, both direct planetary settlement and the development of interplanetary habitats will evolve together to provide a wide range of opportunities for human expansion throughout the solar system. If all this comes to pass and the solar system becomes the new home for man, then we must be prepared for radical changes in the human condition.

Consider, for example, the implications of population growth off Earth. Given the depth of the Earth's "gravity well", and the consequent cost of sending people into space, it seems unlikely that sizeable numbers will ever emigrate from Earth. Certainly, any idea that space migration could, through removing surplus population, stabilize or even lower the Earth's population, should be forgotten, for the cost of removing and settling in space hundreds of millions of people a year would be truly enormous [34].

There would be no need, however to send large numbers of colonists into space in order to develop major populations in the solar system. As the history of the colonization of virgin islands indicates, exponential population growth curves rise steeply when people expand into previously uninhabited regions [35]. Of course some change in attitude and institutions might have to take place lest we take into space the low birth rates of our industrial societies [36]. Assuming that any social or physical obstacles to population growth in space could be overcome, it would not take too many generations for an initial population of a few hundred to expand into the millions, and then later into the thousands of millions. Although models of solar system colonization such as David Criswell's in which hyperadvanced engineering techniques harness the Sun in order to provide energy to support as many as 8×10^{16} people may be extreme [37], it is not difficult to imagine that one day many more people will live off Earth than on our planet.

The solar system economy that would support this population would be immense compared to todays's global economy, for space colonists would have at their disposal the power of the Sun, and the minerals and other elements to be found in abundance on the planets and asteroids. How that economy would be organized is impossible to forecast. Particularly given improvements in information technology, one can imagine all economic activity from the smallest asteroidal settlement, to the largest planetary unit, being organized into one highly unified system, operated according to free market principles, or central direction, or a combination of these.

Alternatively, it is much more interesting to speculate that the solar system economy might be polycentric, with many different types of organization co-existing within a loosely coordinated network of quasi-autonomous units.

For example, what kind of organization might result when workers in a space habitat manufacturing solar power satellites or some other valuable export product, declared thenselves owners of their habitat, and used the proceeds from the sale of their products to pay off the original investors, be they governmental or private? Would this result in a variety of high-technology communism, never dreamed of by Marx, in which the inhabitants shared equally in the ownership of the means of production and the revenues generated? Or, would inequality inevitably reappear as the more technically and manageriably adept assert control and appropriate for themselves more than an equal share of the benefits?

How would a populated solar system be organized politically -as a collection of many independent units, or as a unified polity? Perhaps both possibilities could occur- in sequence. A common theme in science fiction is for space colonies to declare their independence from Earth. If that were to happen frequently, Then we might look forward to a solar system populated by myriads of independent city-states. However, as space colonies multiply in the inner solar system, and particularly if populations there greatly expand, one can imagine to growth of huge new political units wherein individuals colonies lose their independence, and the emergence of new forms of tyranny in the resulting mass societies comprising many thousands of millions of people.

If such a centralization were to occur, there might still be room on the frontier of human settlement for small, separatist communities made up of people who cannot stand centralized political control and mass society. For example, the scenario of space colonization that excites physicist Freeman Dyson [38] is one in which the cost of space travel and the construction of space habitats falls to the point where small groups, such an extended family or a religious cult, could afford to colonize their own asteroid or comet. He sees this possibility as the one hope for humanity of escaping the stultifying effects on the human spirit of life in large and regimented space colonies. Such a safety valve of small group colonization would promote a needed diversity, bringing forth a multiplicity of diverse social experiments from which might emerge brilliant new formulas for organizing society.

Whatever the details of how life will be organized in circum-solar space, it seems clear that the successful colonization of the solar system will lead to a major shift in how we think about ourselves and our relationship with nature. By then we would have realized that we are no longer bound to Earth, and that it is within our power to establish colonies virtually anywhere in space. The world view of Earth-bound humanity will then have been replaced by a truly cosmic view. At this point human consciousness will almost literally be adrift in space, and Arendt's fear that space colonization would lead to a radical change in the human condition will have been realized, at least on the philosophical level.

Interstellar Migration

Seen in this light, the colonization of the solar system would be just the prelude to interstellar migration. By learning how to build viable habitats both on other planets and in interplanetary space, by learning how to make life flourish in these environments, and by learning how to develop cultures adapted to the challenges and rewards of space living, our descendants will have developed the means for establishing human colonies outside the solar system, as well as have adopted the cosmic view required for interstellar migration.

At this point, if not before, the reconnaissance of other star systems to locate and explore environments suitable for implanting human life will be undertaken in earnest. Similarly the search for extraterrestrial intelligence (SETI), by radio observations and perhaps more active methods, will be vastly intensified, for it will be vital to know if we are alone in the galaxy and therefore free to expand [39]. Let us assume, as contemporary theory and preliminary observations would lead us to believe, that a significant proportion of star systems in the galaxy do prove to be hospitable to human life. Let us further assume -although here both theory and observation lags- that any extraterrestrial civilizations which might exist are so widely dispersed that there would be plenty of room in the galaxy for expansion.

That would leave one major problem to be solved: interstellar travel. To reach inhabitable star systems, there must be developed either very fast spaceships which could make interstellar voyages within a human lifetime, or slower interstellar arks in which generations of space voyagers might be nurtured in order to deliver a final, colonizing generation to a distant destination. Here there has been no lack of ingenious plans.

For example, one of the latest ideas for solving the problem of providing fuel for a spaceship employs a variation of the light sail idea. It involves building a solar-fueled power station from materials mined on Mercury, then placing it in close orbit around the Sun, and using it to beam microwaves to push a spaceship equipped with a giant microwave sail towards the nearest star [40, 41 & 42].

As for interstellar arks the ideas of O'Neill about fabricating very large space habitats from lunar materials have led to a number of recent elaborations of this venerable science-fiction theme in which such habitats are sent on multi-generational voyages to other star systems [43 & 44].

Eric Jones and I have suggested an even slower method of interstellar travel: colonize the comets of the Oort cloud and then nudge a comet colony out of its solar orbit and send it on a 50,000 year voyage to another star [42]. While none of these and other plans are anywhere near to being realized, given human ingenuity, and the drive to spread out into space that will intensify as people adapt to space living and begin to look more and more beyond our solar system, it is not difficult to imagine that sometime in the future the problem of interstellar flight may be solved.

Let us again assume that some man-made or natural catastrophe does not extinguish human life, this time in the entire solar system. Then, with the technology of colonization available, and a strongly developed cosmic attitude favoring expansion, there is every reason to believe that interstellar migration will proceed and that humanity will become established among the stars.

To imagine the possible outcome of this migration, let us try to peer millions of years into the future, for if we must look back five million years to see the beginnings of humanity, we should be prepared to look at least that far into the future in order to think seriously about where we are going. For purposes of developing this distant perspective, let us suppose that our descendants have been successful in establishing initial colonies in nearby star systems, and that these colonies in turn have sent out migrants to other star systems, and so on until our region of the galaxy if filled with millions of colonies of Earth-descended societies. In this extreme though logical outcome on man's urge to explore, what would happen to the human conditions?

To be sure, it is utterly impossible to forecast so far into the future. However, by focusing on human evolution we can immediately see one broad theme emerging. If our descendants spread far and wide among the stars, then humanity must diversify, first culturally, then biologically. Diversification would follow separation, for once cettlements become separated by light years, humanity can no longer remain unified.

Consider the case of cultural diversification. Each new colony would in effect become an independent cultural experiment. New and disparate ways of organizing society, of allocating goods and services, and of defining what is just and moral, and so on, would be chosen by conscious design or would develop through random cultural drift or other forces. This diversification, which would increase as humanity spreads and the frontier expands farther and farther outward from Earth, could not be reversed by the organization of galactic empires -unless travel faster than the speed of light would someday be possible, which seems most unlikely. For societies separated by light years, even the ability to travel at some significant fraction of the speed of light would not enable structures of political domination to be maintained over long distances. Nor could information exchange at the speed of light keep widely-separated societies perpetually in step with one another. Even if the colonists tried, the tyranny of light years would make inter-cultural dialogue most difficult. Furthermore, as the number of colonized star systems multiplied the problem for any one unit to keep track of the others would progressively become more and more difficult to the point of impossibility.

But, biological diversification is the real threat to the unity of mankind, for by scattering through space our descendants will promote an explosive speciation [45]. Biologists tell us that speciation occurs most readily in very small populations that have become geographically isolated from the main stock [45 & 46]. While genetic change is resisted by large populations well adapted to their environment, favorable genetic mutations can easily gain a foothold in marginal geographic areas -where often pressures for natural selection can be intense- and then spread quickly through the small populations isolated there.

While this scenario of speciation has probably been repeated several times in our hominid past, for at least the last 50,000 years or so there is no evidence of major genetic change in mankind. The paleontologist Stephan Jay Gould [47] estimates that the average Cro-magnon Man, properly trained, could have handled computers with the best of us, and concludes that all we have accomplished since his day has been the result of cultural evolution. The possibilities for speciation by small, isolated groups inherent in the spread of *Homo sapiens* over the globe have been largely eliminated by the recently developed mobility of mankind which has made the world community into one large population unit in which genes flow back and forth from continent to continent.

The migration of our descendants into interstellar space, and the consequent scattering of small, discrete breeding populations separated by light years from one another will once more set the stage for speciation. In addition, genetic change may be greatly accelerated by the highly ionized radiation found in space, and the selective pressures harsh space environments would impose. Furthermore, advances in genetic engineering forbidden on Earth, may be more readily applied in interstellar space. When the nearest neighbors are light years away, groups desperate to adapt physiologically to an extreme environment, or anxious to enhance the intelligence of their descendants, will feel much freer to experiment and develop their own desired version of humanity than would inhabitants of the closely settled solar system.

So, by setting off in small groups to populate space human evolution would be greatly accelerated. Nevertheless, however exotic this interstellar migration scenario might seem, the process of biological change would echo that followed by the exploratory founders of other species, including those in the hominid line. Typically it is from the adventure some minority, not the main stock, that evolutionary advance flows [48].

Biological change would not be limited to a single speciation event, or a unilinear series of them. Space is not a single environment. It is a geocentric category we use to refer to everything which is not on Earth. There are innumerable environments in space, and with human dispersion through space there will be innumerable opportunities for speciation. By migrating into space we will be embarking on a process of the adaptive radiation of humanity that will spread life as far as technology, or any limits placed by other intelligent life forms in the galaxy, will allow. Just as the movement of fish onton land led to the multiplication of life forms on Earth, so will this movement into interstellar space lead to an explosive radiation of intelligent life into the galaxy. In the words of Freeman Dyson [38], to question whether or not we shall expand into space is really to ask: "Shall we be one species or a million?"

What exact forms will our interstellar descendants take? Will some evolve into creatures which are little more than huge brains? Will others develop thick, armor-like skin to protect themselves against hazardous space conditions? Will some attempt to combine computers and other man-made systems with biological forms [49 & 50]? I do not know, and doubt that our evolutionary future can be predicted beyond the realization that humanity will speciate if we follow our urge to expand into space. Despite the fact that we have labeled our own species as <u>sapiens</u> and are beginning to be aware that we are not the culmination of all evolutionary progress, we are probably in no better position to predict the exact course of our future development than were those first hominids able to even roughly foretell theirs. If five million years ago they could not have even dimly conceived the evolutionary consequences of their move from the forest to the grasslands, how can we -despite all our knowledge- forecast exactly what evolutionary developments will follow over the next five million years as our descendants disperse among the stars?

References

[1] Arendt, H., 1958, The Human Condition, University of Chicago Press, Chicago

[2] Sevastyanov, V., A.D. Ursul and Yu. Shkolenko 1981, The Universe and Civilization, (Translated from the Russian), Progress Publishers, Moscow

[3] Darwin, C., 1871, The Descent of Man and Selection in Relation to Sex, John Murray, London.

[4] Johanson, D., and White, T., 1979 A Systematic Assessment of Early African Hominids, Science. 205, n° 4378, p. 321-324

[5] Sarich, V.M. and A.C. Wilson, 1967, ¢ An Immunological Time-Scale for Hominid Evolution, Science, 158, n° 3805, p. 1200-1203

[6] Goodmann, M., and R.E. Tashian, 1976, Molecular Anthropology, Plenum, New York

[07] Yunis, J.J. and O. Prakash, 1982, The Origin of Man: A Chromosomal Pictorial Heritage, Science, 215, n° 4359, p. 1525-2529

[8] Johanson, D., and Edey, M., 1981, Lucy: The biginnings of Humankind, Simon and Shuster, New York

[9] Tanner, N.M., 1981, On Becoming Human, Cambridge University Press, Cambridge

[10] Campbell, B.G., 1982, Humankind Emerging, Little, Brown, Boston

[11] Leakey, L.S.B., P.V. Tobias and J.R. Napier, 1964, A New Species of the Genus Homo from Oldavai Gorge, Nature, 202, n° 4297, p. 7-9

[12] Kalb, J.E., C.J. Jolly and E. Mebrate, 1982, Fossil Mammals and Artefacts from the Middle Awash Valley, Ethiopia, Nature, 298, n° 5869, p. 25-29

[13] Washburn, S.L., 1967, Behaviour and the Origin of Man. Huxley Memorial Lecture, Proceedings of the Royal Anthropological Institute, p. 21-27

[14] Cann, R.L., M. Stoneking, and A.C. Wilson, 1987, Mitochondrial DNA and Human Evolution, Nature, 325, p. 31-36

[15] Allen, J., J. Golson and R. Jones, 1977, Sunda and Sahul, Academic Press, London

[16] Reeves, B.O.K., 1983, Bergs, Barriers and Beringia: Reflections on the Peopling of the New World. In Quaternary Coastilines and Marine Archeology, P.M. Masters and N. C. Flemming (eds), Academic Press, London, p. 389-412

[17] Birdsell, J.B., 1957, Some Population Problems Involving Pleistocene Man. Cold Springs Harbor Symposium on Quantitative Biology, 22, p. 47-69

[18] Finney, B.R., 1977, Voyaging Canoes and the Settlement of Polynesia, Science, 196, n° 4296, p. 1277-1285

[19] Finney, B.R., 1985, Voyagers into Ocean Space. In Interstellar Migration and the Human Experience, (B.R. Finney and E.M. Jones eds.), University of California Press, Berkeley, p. 164-179

[20] Parry, J.H., 1974, The Discovery of the Sea, Dial, New York

[21] Godinho, V.M., 1965, Os Descobrimentos e a Economia Mundial, Arcadia, Lisbon
[22] Wallerstein, I., 1974, The Modern World-System, Academic Press, New York
[23] Dobzhansky, T., 1962, Mankind Evolving, Yale University Press, New Haven
[24] Baker, R., 1980, The Mystery of Migration, Macdonald, London
[25] Gould, S.J., 1977, Ontogeny and Phylogeny, Harvard University Press, Cambridge
[26] Eiseley, L., 1957, The Immense Journey, Franklin Watts, New York
[27] Lorenz, K. 1971, Studies in Animal and Human Baheviour, 2, Harvard, University Press, Cambridge
[28] Ehrlich, P.R., C. Sagan, D. Kennedy and W.O. Roberts, 1984, The Cold and the Dark, W.W. Norton, New York
[29] Ehricke, K.A., In Press, The Seventh Continent: Industrialization and Settlement of the Moon, Krafft Ehricke Institute, La Jolla, California
[30] Hartman, W.K., R. Miller and P. Lee, 1984, Out of the Cradle, Workman, New York
[31] Lovelock, J. and M. Allaby, 1984, The Greening of Mars, Warner Books, New York
[32] McKay, C.P., 1985, The case for Mars II, 62, Science and Technology Series, a Supplement to Advances in Aeronautical Sciences, American Astronautical Society, San Diego
[33] O'Neill, G., 1977, The High Frontier, Morrow, New York
[34] Hardin, G., 1978, Stalking the Wild Taboo, Kaufman, Los Altos, California
[35] Birdsell, J.B., 1985, Biological Dimensions of Small, Human Founding Populations in Interstellar Migration and the Human Experience, B.R. Finney and E.M. Jones (eds), University of California Press, Berkeley, p. 110-119
[36] Finney, B.R., In Press, Anthropology and the Humanization of Space, Acta Astronautica.
[37] Criswell, D.R., 1985, Solar System Industrialization: Implications for Interstellar Migrations. In Interstellar Migration and the Human Experience, B.R. Finney and E.M. Jones (eds.), University of California Press, Berkeley, p. 50-87
[38] Dyson, F., 1979, Disturbing the Universe, Harper, New York
[39] Finney, B.R., 1985, SETI and Interstellar Migration, Journal of the British Interplanetary Society. 38, p. 274-275
[40] Dyson, F., 1982, Interstellar Propulsion Systems, In Extraterrestrials: Where are they? M.H. Hart and B.. Zuckerman (eds.), Pergamon Press, New York, p. 41-45
[41] Forward, R.L., 1982, Roundtrip Interstellar Travel Using Laser-Pushed Lightsails, Hugues Research Laboratories Research Report n° 550
[42] Jones, E.M., and B.R. Finney, 1985, Fastships and Nomads: Two Roads to the Stars. In Interstellar Migration and the Human Experience, B.R. Finney and E.M. Jones (eds.), University of California Press, Berkeley, p. 88-104

[43] Bond, A., and A.R. Martin, 1984, World Ships An Assessment of the Engineering Feasibility, Journal of the British Interplanetary Society, 37, p. 254-266
[44] Martin, A.R., 1984, World Ships Concept, Cause, Cost, Construction and Colonization, Journal of the British Interplanetary Society. 37, p. 243-253
[45] Valentine, J.W., 1985, The Origins of Evolutionary Novelty and Galactic Colonization, In Interstellar Migration and the Human Experience, B.R. Finney and E.M. Jones (eds), University of California Press, Berkeley, p. 266-277
[46] Mayr, E., 1954, Change of Genetic Environment and Evolution, In Evolution as Process, J.S. Huxley, A.C. Hardy and E.B. Ford (eds.), Allen and Unwin, London, p. 157-180
[47] Gould, S.J., 1982, The Panda's Thumb, Norton, New York
[48] Stanley, S.M., 1979, Macroevolution: Pattern and Process, W.H. Freeman, San Francisco
[49] Drexler, E., 1986, Engines of Creation, Anchor Press/Doubleday, Garden City, New York
[50] Jastrow, R., 1981, The Enchanted Loom, Simon and Shuster, New York

Interventions

Jean-Pierre FAYE

Vous définissez l'homme comme l'être de migration, le migrant. C'est la plus belle des définitions et la plus pertinente -surtout si l'on ajoute cette composante : il est celui qui rapporte, dans le langage, les effets de migration.

Or c'est de cet effet de migration que l'on peut peut attendre de grandes surprises.

Le lieu de la plus grande condensation des migrations a été la Mésopotamie. Et c'est là qu'à la migration sumerienne, se superpose la migration akkadienne. Au peuple qui a inventé l'écriture pictographique vient s'ajouter celui qui va, en conquérant, dicter aux scribes sumériens une parole sans écriture : ainsi s'invente l'écriture phonétique, capable de fixer le trait phonologique. Loin d'être une chute ou un "abaissement" de l'écriture, comme ce fut la mode fantasmatique de l'affirmer durant la précédente décennie, c'est là l'essor le plus grand de la pensée, qui s'est opéré à la jonction migratoire d'une langue dite sémitique et d'une langue dite agglutinante. Essor qui rend tout possible, de ce qui sera l'histoire humaine en son sens le plus vif.

La question qui se pose, et que déjà, pour conclure, j'aimerais énoncer avec vous, serait : quel effet mental peut-on voir survenir, à partir des migrations hors-Terre ? Quel choc sur le rapport social, sur le langage, sur la pensée...

Gérard Huber

- Vous avez rendu hommage aux Africains et aux Polynésiens dans votre exposé. Lorsqu'on s'aperçoit du destin, que la civilisation occidentale leur a assigné, on peut se poser la question : pourquoi nous enthousiasmons-nous pour la colonisation de l'espace ?

- Vous construisez une idéalité de l'espèce future vivant dans l'espace autour de l'espoir d'une fin de la peur et de la nostalgie. On peut se demander ce que cela implique pour la vie quotidienne de notre humanité aujourd'hui.

STATE OF OUTER SPACE

Lubos Perek
Astronomical Institute
Czechoslovak Academy of Sciences, Prague, Czechoslovakia

1. Introduction

At this colloquium we are discussing the building of large structures in outer space capable of accommodating hundreds of people, possibly thousands. We are not building these structures now because the technology and the finances are not available, yet we might be, through our other space activities, deciding about the possibility to build such structures at all, or at least about the difficulties which will be encountered during the construction phase.

The first space stations, such as those which are at present on the drawing boards or partly already in space, will be located, we may safely assume, in low Earth orbits. There will have to be frequent contacts with the ground, such as receiving supplies of materials needed for sustenance of life, for scientific application programmes and for further construction and extension of the station. Also the crew will live, in that phase, on the station for a limited period of time and will have to be relieved after the termination of their term of duty.

During all that time the station will be exposed to the hazards of the natural, as well as artificial environment at an altitude of a few or several hundreds of kilometres.

In the next phase, requiring a certain degree of autonomy, the station will possess a certain degree of manoeuvering capability and could be located at a high Earth orbit providing a very long lifetime, or close to the Moon, or Mars, or one of the asteroids for the exploitation of materials found on these celestial bodies. Even during that phase the contacts with the Earth could be frequent and indispensable. Thus a safe passage through, or a stay in, low Earth orbit will be necessary for a long period of time.

One of the important aspects of a long existence of a space station or space settlement is its safety. Not only its internal safety, involving the perfect functioning of all its parts and systems, but also its external safety which depends on the quality of the environment, in particular on the numbers, sizes and velocities of other objects orbiting in the vicinity of the station. What do we know about such objects and about the dangers of close approaches or collisions ?

J. Schneider and M. Léger-Orine (eds.), Frontiers and Space Conquest, 173–182.
© 1988 by Kluwer Academic Publishers.

2. Man-made Space Objects

There is nothing we can do about the population of natural objects which enter the near-Earth environment from interplanetary space. It has been estimated that 2 to 6 x 107 kg of meteoritic material enter the atmosphere per year. But this amount is equalled, or exceeded, by the man-made material orbiting in space, for the greater part in the form of abandoned satellites, large and small debris, down to particles generated by rocket engines. And there are ways how to restrict this man-made population of space objects and thus to make all the future grandiose projects safer, less expensive and easier to build.

There are basically two kinds of man-made space objects. Those larger than about 4 cm in low orbit, or larger than 10 cm at 1000 km altitude, can be detected by present tracking networks. These are the trackable space objects. Smaller objects, in particular at high altitudes, are called non-trackable objects.

One of the publicly accessible sources of information on trackable space objects is the NASA Satellite situation report [1] which is based on a network operated by the North American Aerospace Defense Command. Since only objects which have been observed by more than one radar and which could be associated with a launch are included, the list may be incomplete in the number of debris, especially of smaller sizes.

According to [1] there were as of 30 septembre 1986:

- 1635 payloads in orbit,
- 4476 debris in orbit, while
- 1887 payloads and
- 9004 debris have already decayed.

It is estimated [2] that of the total population of trackable space objects in orbit are:

- 5 % operational payloads,
- 21 % non-operational payloads, i.e. satellites which terminated their function,
- 25 % mission-related debris and
- 49 % debris from satellite break-ups

The number of trackable space objects has been increasing since the beginning of the space era as shown in figure 1. The only exception were the years 1978-1982 when an abnormal number or space objects decayed due to the maximum of solar activity. During that period the density of the atmosphere at high altitude is considerably higher than during a solar minimum. The factor of increase is, on the average, 10 at altitude 800 km, 20 at 500 km and 4 at 300 km. Since the density of the atmosphere is directly responsible the atmospheric drag which slows down orbiting space objects, their lifetimes get shorter.

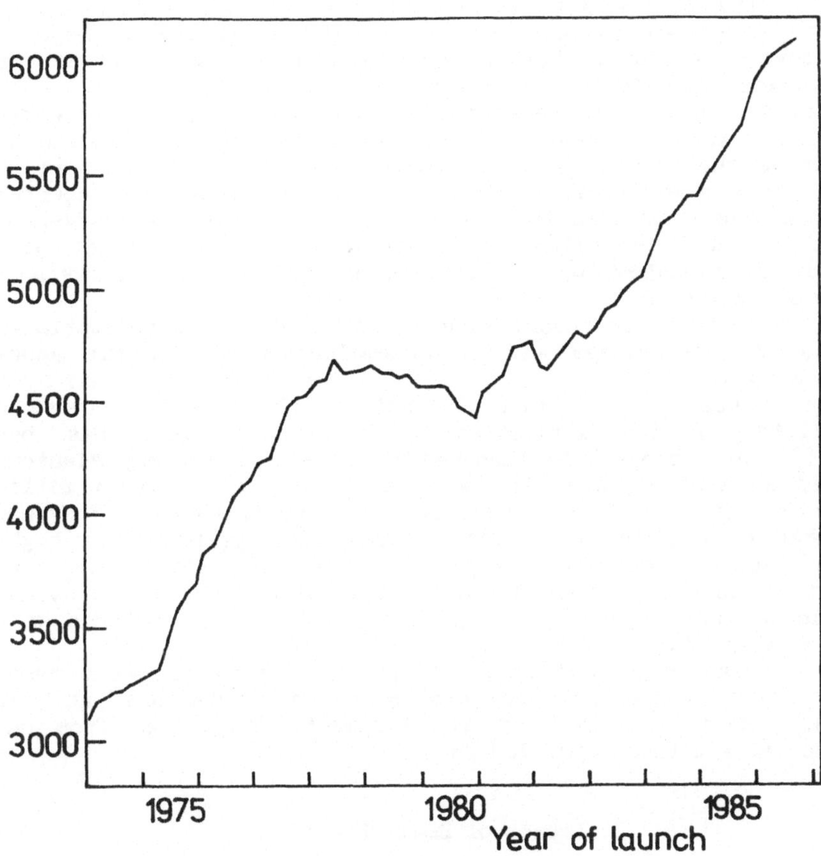

Figure n° 1
Number of trackable space objects

The amount of non-trackable debris is unknown. It has been estimated that only one third of all objects larger than 4 cm has indeed been detected. For smaller debris, down to 1 cm size, the inclompleteness of detection is certainly much higher.

Some of the debris are related to the normal function of the spacecraft. These are called mission-related. These are the rejected shrouds, covers, explosive bolts. On the average, their numbers are small and their sizes tend to be large.

Most of the debris are related to explosions. The first break-up occurred in 1961. It created 271 debris out of which one fourth has decayed while the remaining three fourth ar still in orbit. In the following years some 80 additional explosions or break-ups occurred. Some of these were related to propulsion. The cause was inadvertent mixing of residual propellants left on board of the rocket body. Since 1981 a procedure was instituted with the Delta rockets to prevent such explosions.

Other explosions occurred during injection into the geostationary orbit. One of the causes may be anormalous burns in the apogee injection motors.

Another cause of explosions are antisatellite weapons tests or other military-related activities. One recent test has been reported [3] on a projectile launched to intercept a dummy Minuteman warhead at an altitude over 160 km. At impact, more than a million fragments were generated and strewn over an area of 40 km^2. They were photographed by tracking telescopes. Such tests performed at higher altitudes would generate large numbers of long-living and non-trackable debris. These intentional explosions could be avoided -it should be noted -just by changing the intention. No advance or expense in technology is required.

Yet, the causes of one half of the explosions in orbit remain unkown. Most dangerous are explosions or break-ups at high altitudes. More than 60 % of all fragments originated from only 9 break-ups at altitudes over 700 km.

3. Impact of Small Debris

The impact of even a small debris on the structure of a spacecraft can cause important damage. Tests on hypervelocity impacts into structures similar to those used on spacecraft have shown that during a collision of a small object with a relatively large structure, the small impacting object will melt or evaporate and form a number of small high velocity fragments. The large object will develop a crater or hole with a possible extensive fragmentation. Since the relative velocity is large, on the average of about 10 km/s, the mass ejected from the large object will be more than 100 times the mass of the small object.

The presence of small particles and flakes has been confirmed by observation. One half of the hypervelocity pits on windows of the Apollo Command Module were aluminum lined. Another example was observed during the flight of STS-7, 1983-59A. The crew detected a pit on the outside window of about 5mm in diameter. When examined after landing, the pit was shown to contain titanium with a trace of aluminum. The impacting debris had a diameter of 0.2 mm and impacted at a velocity of 3-6 km/s. It could have been a flake of paint from another space object.

4. Collision Probability in Low Orbits

There are methods in physics to compute the probability of collision if the number of objects in a given volume is known, as well as the average relative velocity of two objects and the average cross-section of the target.

The problem of collisions in space is more complicated because of several factors. In the first place, there is a wide range in sizes of space objects, in the second place, the number of debris increases after each collision and, last but certainly not least, there are crowding effects in space which may result in a collision probability considerably exceeding the average value derived from a uniform distribution of debris in space.

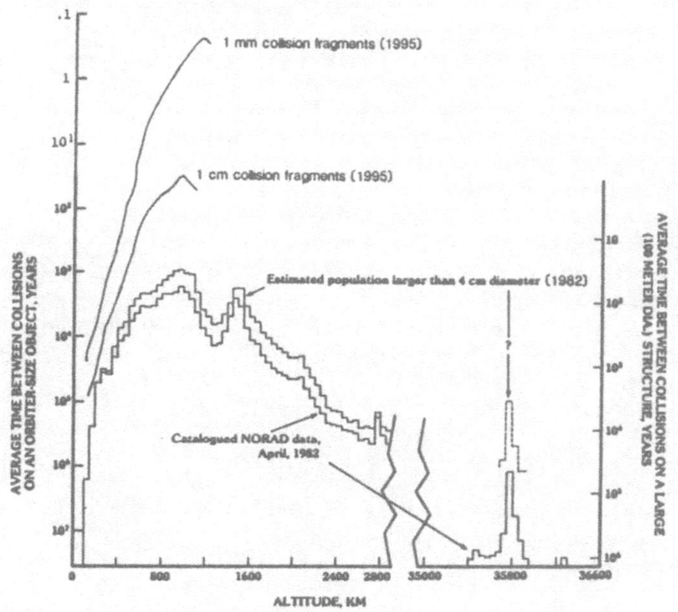

Figure n° 2

Average time between collisions for the current population and for the population of fragments as predicted for 1995

Figure 2 shows the computed average collision probabilities according to [2] for the trackable population, for the extimated population of objects larger than 4 cm and for a population of smaller objects as they are estimated to exist by 1995. The dependence of collision frequencies on the size of the target is illustrated by the vertical scale at right. It refers to a space station of diameter 100 m.

A few collisions might have already occurred in space although it is extremely difficult, if not outright impossible, to furnish a conclusive proof. Some malfunctions of satellites have, however, already been observed without a better explanation ever having been proposed. This is the case of GEOS 2, 1978-71A, which developed a fault three days after it became fully operational. There might have been a mechanical damage to a small section of one solar array panel. Another case was Cosmos 954, 1977-90A, which was in normal operation until 6 January 1978 when it lost pressurization, started to tumble and rapidly decayed. Its fragments, some of them from its nuclear power source, were collected in northern Canada. The opinion was expressed by L. Sedov [4] that the satellite collided in flight with some other object of natural or artificial origin. As a third example, let us mention the break-up of Pageos, 1966-56A, which was in an almost polar orbit at an altitude over 4000 km. It generated large numbers of fragments on two occasions, in 1975 and 1976.

The situation is somewhat different in the geostationary orbit. There are at present some 280 objects, at least 100 of them either inactive or non-functional. The difficult point is the number of non-trackable objects because at the distance of 36,000 km even objects of 1 m size escape detection and nothing is known about their number. The collision probability is still small -it is measured by an interval of hundreds of years between two collisions- but the risk of losing a valuable satellite is real. Some agencies, e.g. the Air-Force Satellite Control Facility [5], are monitoring close approaches between selected communication satellites and all trackable objects. An alert is sounded when the predicted approach is below 50 km and a collision avoidance manoeuvre is considered at 5-8 km separation and is implemented if the accuracy of positions is verified. Computations made for a sample of 21 satellites over a six month period in 1981 showed that there were 120 encounters below 50 km. Several approaches were within the 1-5 km range and in these cases collision avoidance manoeuvres were made. Due to crowding at some longitudes, the probabilities of collision were up to two orders of magnitude greater than those computed from the average density of objects in the geostationary orbit.

5. Hazards to Peaceful Space Activities

Hazards to peaceful space activities are posed by objects in space which have either no use at all or which have a use contrary to peaceful purposes.

Among the objects which have no use are not only inactive

satellites which terminated their functions and the debris, but also some space missions with questionable aims. As an example, a project called URNSAT was reported some time ago in the press. It proposed to launch some ten thousands vials containg ashes of deceased people into a high Earth orbit with a very long lifetime. The idea perhaps was to provide an undisturbed resting place. In practice, a collision or malfunction might lead to a break-up of the spacecraft and the resting place might turn into a threat to manned as well as unmanned space missions. What a frightening perspective for possible customers !

Also missions launched with a view to provide a symbol on the night sky should incorporate a suitable provision for terminating the mission at a time when the symbole could turn into an undesirable burden.

Missions contrary to peaceful purposes are of course all large military projects, in particular those which require putting weapons into space. Even if such a system is never used, it has to be tested by intentional explosions of space objects on an extensive scale covering long period of time. Such projects, if tested or deployed, would require a re-evaluation of the collision probabilities as derived above.

On a more cheerful note, one of very important possible hazards was prevented by international agreements. It is the co-ordination of radio communications to and from space objects. Available transmitting frequencies are a scarce resource and requirements are high. The co-ordination is performed within the framework of the International Telecommunication Union. The procedures are highly complex but it is an admirable piece of work, at the same time preventing harmful interference in communications and respecting the principle of national sovereignty.

One more problem should be mentioned in this context. For restricting collisions or close encounters it is necessary to know positions and motions of all trackable space objects. This knowledge comes in the form of complete and updated sets of numerical values of orbital elements. Such information is at present obtainable only from national sources -with the exception of classified satellites. International sources, such as the United Nations Central Register, or lists provided by non-governmental organizations, give only three or four of the necessary six orbital elements for each object and even these elements are updated only on rare occasions.

6. Reducing Hazards to Safety in Space

Any steps which would efficiently reduce safety hazards would have to be taken on an international level, in co-operation between all present and potential future launching countries. The only global forum which has a tradition in this respect is the United Nations.

The United Nations became interested in space activities right after the beginning of the space era. A Committee on the Peaceful Uses of Outer Space was established in 1959 and after a few years of deliberations elaborated a Declaration of Legal Principles Governing the Activities of States in the Exploration and Use of Outer Space [6]. Many of the principles adopted have a high ethical value: Outer space and celestial bodies are free for exploration and use by all states. Outer space and celestial bodies are not subject to national appropriation by claim of sovereignty. Space activities shall be conducted in the interest of maintaining international peace and security and promoting international co-operation and understanding. States bear international responsibility for national activities in outer space and shall be guided by the principle of co-operation and mutual assistance. States are internationally liable for damages caused by space objects. States shall regard astronauts as envoys of mankind in outer space and shall render to them all possible assistance in case of need.

These principles were transformed into international law in the 1967 Outer Space Treaty [6]. The treaty contains additional provisions prohibiting the placing of nuclear weapons or weapons of mass destruction in orbit or on celestial bodies. The establishment of military bases or testing of weapons is prohibited on celestial bodies. Several of the principles have been elaborated in more detail in subsequently adopted instruments of international law.

The 1979 Moon Treaty contains provisions reserving celestial bodies -but not near-Earth orbits- exclusively for peaceful purposes. It also prohibits the disruption of Moon's environment. Regrettably, the Moon Treaty has been ratified, as of now, by only five countries, major spacefaring countries not included.

Thus, a body of internationally recognized law came into existence, incorporating with foresight many principles of permanent validity.

The present state of space activities can be characterized by a much higher density of space traffic, as compared to the early times when it was tacitly assumed that a space mission is, if not quite alone in space, at least independent of other missions. The fact that space missions are not independent but rather can be influenced by other missions, requires the adoption of additional principles into space law. Moreover, ground has to be prepared for the projects of the future, for space stations and settlements. Among the first steps, the following would deserve consideration:

1. Declaring outer space to be reserved exclusively for peaceful purposes. This reservation applies, according to the Outer Space Treaty, to the Moon and other celestial bodies. In the Moon Treaty it has been extended to orbits and trajectories to and around celestial bodies but it does not apply to orbits around the Earth.

2. Defining the limit of outer space. Such a limit, conveniently expressed in terms of altitude above the sea level, e.g. 100 km, would determine the region of validity of space law. Remarkably, none of the instruments of space law contains such a definition.

3. Strengthening existing provisions on availability or orbital data of objects launched into outer space. An internationally agreed mechanism should be adopted for making avaliable complete and updated sets of orbital elements in order to increase safety of manned as well as unmanned missions.

4. General provisions should be adopted, again on an international level, for the protection of near-Earth outer space as part of the Earth's environment. Several partial measures have been made within the framework of various programmes or treaties but there is no certainty that all aspects have been covered or that all relevant effects are being monitored.

5. A quite indispensable step is the adoption of an agreement on some sort of Traffic Rules for outer space, or of a Code of Conduct in space, containing provisions or recommendations for restricting the amount of debris, for removing inactive satellites from orbital belts used by active systems, in general for all practices restricting the probability of collisions.

These items and probably many others are urgently needed now to facilitate the building of space stations in the near future and to make possible the settlement of space in far perspectives.

References

More details on the above subjects can be found in :

- L. Perek
  Traffic Rules for Outer Space, Proc. 25th Coll. on the Law of Outer Space, held in Paris, 1982, AIAA New York 1983, p. 37

- L. Perek
  The Environmental Impact of Space Activities, to appear in Outer Space -A Source of Conflict, (eds.), B. Jasani, a study prepared by the Stockholm International Peace Research Institute and the United Nations University

[1] Sattelitte situation report, NASA, Office of Public Affairs, Goddard Space Flight Center, last issue, 26, n° 3, September 30, 1986

[2] D.J. Kessler
Orbital Debris Issues,, Presented at the COSPAR Congress, Graz, Austria, 1984.

[3] H.A. Bethe, R.L. Garwin, K. Gottfried & H.W. Kendall
Space-Based Ballistic-Missile Defense, Scientific American, 251, n° 4, p. 37, 1984.

[4] L. Sedov
quoted in Spaceflight, 20, p. 184, 5 May 1978

[5] M.G. Wolfe, V.A. Chototov, E.E. Bond
Man-made Space Debris -Implications for the Future, Space Safety and Rescue 1982-1983, 58, Science and Technology Series, Amer. Astronaut. Soc. Publi., p. 43

[6] Space Activities of the United Nations and International Organizations, United Nations, New York, 1986, Document A/AC.105/358.

Intervention

Jeff Hoffman

- NASA is developing an optical tracking network to give more sensitivity to small space debris;

- the two communications satellites recovered by the space shuttle in November 84 showed evidence of impact after about 9 months in space;

- when LDEF is finally recovered and returned to earth, we will have much more information on the small end of the spectrum of space debris, both natural and artificial.

UN EXEMPLE D'EVENEMENT SPATIAL : LA TOUR EIFFEL DE L'ESPACE

Philippe Gillieron
Tour Eiffel
Champ de Mars
75007 Paris, France

Louis Leprince Ringuet, de l'Académie Française, ancien professeur de physique atomique ici-même, à l'Ecole Polytechnique, était le 24 novembre dernier aux côtés de Bernard Rocher, le Président de la Société Nouvelle d'Exploitation de la Tour Eiffel, et d'André Lebeau pour désiger le lauréat du concours de la Tour Eiffel de l'Espace parmi trois projets.

Au préalable, le 10 octobre précédent, sous la présidence d'André Lebeau et dans le cadre du Congrès de la Fédération Astronautique Internationale à Innsbrück, un jury international avait sélectionné ces trois projets parmi 99 dossiers présentés à ce concours d'idées.

Comment peut-il se faire qu'un projet, dont le prix de revient n'atteint pas le millième des dépenses annuelles consacrées à l'espace à l'échelle mondiale, qui ne présente ni intérêt militaire ni intérêt commercial, puisse voir se pencher sur son berceau autant de fées compétentes et faire l'objet à la seule étape du concours d'idées d'autant de commentaires au plan international ?

Je vous propose, avant de tenter de répondre avec vous à cette question, d'examiner successivement la naissance de l'idée, le déroulement du concours et les conditions de la réalisation du projet.

Reportons-nous tout d'abord cent ans en arrière... Nous sommes le 15 janvier 1887. Dans onze jours exactement va être donné, sur le Champ de Mars à Paris, le premier coup de pioche de la Tour de l'Exposition Universelle de 1889, bâtiment pour lequel a été choisi le projet de Monsieur Eiffel.

Nous sommes réunis, en séance d'académie, ou de société savante et nous débattons d'un sujet brûlant qui va mettre la philosophie à l'épreuve. Lequel ? Sans aucun doute, "la Conquête de la Terre par le chemin de Fer". Un ingénieur et entrepreneur confirmé, Gustave Eiffel, présente à cette séance une communication sur son projet de Tour. Sa place est bien ici car depuis qu'il est sorti de l'Ecole Centrale, il a construit des ponts de chemin de fer, de plus en plus longs, de plus en plus hauts. Et il présente un dessin de son bureau d'études daté de 1884 qui montre ce que serait la structure métallique d'un pylone de 1 000 pieds. Il explique comment cette Tour, conçue comme les piles de viaducs va, par sa hauteur nouvelle,

J. Schneider and M. Léger-Orine (eds.), Frontiers and Space Conquest, 183–187.
© 1988 by Kluwer Academic Publishers.

au-dessus de Paris, faire découvrir aux hommes une nouvelle frontière, bien au-delà de l'aventure du Chemin de Fer dont elle est issue.

L'histoire est à peine imaginaire. En ces temps là, où l'aventure industrielle était à son départ, la Tour de Monsieur Eiffel allait révéler au grand public de l'époque l'apparition d'un monde nouveau, notre monde moderne. Il ne s'agissait d'ailleurs que d'une révélation symbolique, car l'objet était gratuit, ses prétentions n'étaient pas autres que de distraire les visiteurs en leur faisant découvrir la fascination du vide.

Et pendant ce même temps, Jules Verne écrivait. En 1886, on retrouve cette phrase : "on foulera l'air comme on foule la terre." Dans son roman "De la Terre à la Lune", une fusée décollait de Floride.

C'est ainsi qu'est née la Tour Eiffel de l'Espace, en interrogeant Gustave Eiffel sur le geste qu'il ferait, l'objet qu'il proposerait, cent ans après la Tour Eiffel.

L'objet se définissait tout naturellement. De la conquête du Chemin de Fer, l'on passait à la conquête de l'Espace, de l'esprit de clocher à la conscience universelle soutenue par un fantastique développement planétaire de la communication par l'image, d'une industrie française à une industrie européenne.

Dans le respect du processus historique, la Tour Eiffel lançait le 12 juin 1986, anniversaire du jour où Gustave Eiffel gagnait le concours de la Tour de 300 mètres, un concours d'idées avec les conseils techniques du C.N.E.S. et en collaboration avec l'E.S.A.

Les candidats, limités aux ressortissants des pays membres de l'E.S.A., devaient proposer un objet spatial, visible depuis la terre, depuis le maximum de points de notre planète et symbolique de la communication universelle. Il leur était demandé, en outre, de donner des indications de coût et de faisabilité à l'horizon 1989, année du centenaire de la Tour Eiffel. Enfin, il s'agissait aussi de proposer des idées d'applications de l'objet dans le domaine scientifique.

Parmi les quatre-vingt-dix-neuf dossiers présentés au Jury International d'Innsbrück, une trentaine environ pouvait être considérée comme proposant des projets bien étudiés et intéressants, même si certains d'entre eux n'étaient pas réalisables à l'horizon 1989.

A ce point, le Jury décidait d'accorder en plus des trois premiers prix, des mentions qui pouvaient également récompenser des projets non autorisés à concourir, tels les projets proposés par quelques équipes américaines. Neuf mentions étaient attribuées.

Quels sont ces projets ?

Les équipes concurrentes ont résolu de manière à peu près similaire le problème de la visibilité et le problème de la vision depuis tous les points de la terre.

La visibilité est obtenue à partir de grandes surfaces réfléchissant la lumière solaire et l'universalité à partir d'orbites polaires plus ou moins proches de la terre, certains projets utilisant même l'orbite géostationnaire.

Les contraintes de calendrier de lancement imposant pratiquement le lancement en piggy-back, les projets font presque tous usage de techniques de développement dans l'espace de structures très légères contenues au départ dans une enveloppe de volume et poids réduits, compatible avec les volumes résiduels laissés dans le lanceur.

Le lanceur proposé est ARIANE, quelques projets évoquent VIKING ou MARIANE.

Ce qui distingue donc essentiellement les projets du point de vue technique, c'est la manière selon laquelle est obtenue la surface réfléchissante de grande dimension :

. le développement de structures essentiellement gonflabes,
. le développement de structures filaires reliant des panneaux de films réfléchissants ou des volumes gonflables réfléchissants,
. ou la combinaison de ces deux techniques.

Les projets se différencient aussi par l'existence ou non d'un pilotage pour obtenir les effets désirés ; pilotage le plus souvent sommaire, mais faisant appel parfois à des forces d'appui originales comme la force magnétique terrestre ou la pression du flux lumineux solaire.

La variété et l'imagination des approches symboliques sont peut-être plus intéressantes encore.

Quelques candidats ont cherché à retrouver dans le ciel la forme de la Tour Eiffel, mais la plupart d'entre eux ont compris que l'oeuvre symbolique ne visait pas à promouvoir la Tour Eiffel mais bien à renouveler le geste de Gustave Eiffel et donc à symboliser la nouvelle conquête spatiale et la chance qu'elle peut donner au XXIème siècle à la communication universelle.

Ainsi dans les trois projets gagnants, voit-on apparaître l'anneau de l'unité, l'étoile humaine ou le phare de l'espace lançant un message d'espérance.

Mais aussi, dans les projets mentionnés, ce projet anglais de chronomètre de l'espace fait de trois gigantesques aiguilles brillantes donnant au monde entier la même heure de référence.

Ou ce projet américain proposant de résoudre les divergences humaines en donnant à chacun de nous le même angle de vision, c'est-à-dire la vue de la Terre sur un immense miroir gonflable convexe de 203 km de diamètre, se levant chaque jour sur toutes les contrées du monde.

Ou bien ce projet charmant proposant d'extraire les maux de l'humanité symbolisés par la pomme et d'envoyer cette pomme à l'aide d'une voile à propulsion solaire jusqu'à l'un des points de LAGRANGE, où elle se trouvera définitivement immobilisée entre la Terre et la Lune.

L'ensemble de ces projets seront présentés au public pour la première fois à la Cité des Sciences et de l'Industrie de la Villette au mois de mars prochain.

Quels sont les deux projets restant encore en concurrence pour la réalisation éventuelle en 1989 ?

Il s'agit :

- de l'Anneau lumière, le projet de Jean-Pierre Pommereau et de Jérôme Gerber au sein d'une équipe de jeunes architectes. Ce projet est arrivé en tête et a été désigné Lauréat par le Jury de Paris,

- Et de l'Etoile ARSAT de Pierre Comte et Christian Marchal, le projet jugé second, mais restant en course en attendant la sanction des études approfondies de faisabilité et de coût, qui sont menées sur les deux projets.

L'Anneau lumière, cercle constitué de cent sphères réfléchissantes, reliées par une structure filaire en kevlar et mylar de 24 km de circonférence apparaissant dans le ciel comme un collier d'étoiles de magnitude + 1 et traversant l'horizon en dix minutes environ. Le projet, à 800 km d'altitude, aurait sa plus grande dimension apparente voisine de celle de la lune. Sa durée de vie serait au maximum de trois ans. Un dispositif permettrait de rompre la structure filaire et de libérer les ballons à tout moment.

L'Etoile ARSAT est une voile carrée de 60 m de diagonale organisée autour d'une structure de mâts gonflables et possédant un système de pilotage appuyé sur le champ magnétique terrestre qui mettra la voile en position réfléchissante maximum au passage sur un pays et pouvant conduire à une magnitude de l'ordre de - 5,5. En utilisant l'orbite de transfert proposée, les passages sont, à l'apogée, de l'ordre de 20 minutes, avec une période d'intensité maximum de 5 minutes et des possibilités d'éclairs rytmés à partir du pilotage.

Toutes ces données sont actuellement en cours de vérification ainsi que la compatibilité des deux projets avec le planning et les lanceurs.

On peut affirmer aujourd'hui que l'intérêt manifesté par la presse, les télévisions et l'opinion publique autour du projet de Tour Eiffel de l'Espace entraîne la Société Nouvelle d'Exploitation de la Tour Eiffel à étudier, au-delà du concours d'idées, les conditions détaillées de la réalisation.

Bien sûr, des contradicteurs se manifestent. Mais Gustave Eiffel n'a-t-il pas eu à affronter à la fois les riverains du Champ de Mars, qui craignaient l'écroulement du monument, et les grands artistes de l'époque, qui ont adressé au Préfet de Paris une vigoureuse et célèbre protestation indignée ?

L'expérience montre que l'être humain a besoin de symboles, qui révèlent les pensées et les désirs latents puis qui servent à éclairer sa démarche permanente en vue d'assimiler intellectuellement et psychologiquement les conséquences du progès technique.

De même que le programme SPOT donne aux hommes une nouvelle vision de leur Terre, vu de l'Espace, de même la Tour Eiffel de l'Espace peut rendre sensible, en s'adressant à la sensibilité instinctive, l'Espace vu de la Terre et contribuer à une prise de conscience très concrète des développements de la conquête spatiale.

LA CONQUETE INFLECHIE

Daniel Sibony
Psychanalyste, Mathématicien
18, rue du Dragon 75006 Paris, France

Conquête spatiale... L'homme conquis par l'espace, possédé, hanté, soulevé par l'espace -cette fois cosmique. Du reste, y a-t-il d'autre conquête que d'**espace** ? Conquérir une idée c'est d'abord conquérir un espace de pensée, d'être à penser.

Et que nous revient-il de cet élan, de cet élancement dans le vide où se prélève toute forme d'espace ?

D'abord quelques secousses à des notions, des intuitions trop acquises, curieusement, c'est l'expérience ultra moderne qui nous ressource sur l'archaïque, l'originel, et de façon très originale. Certaines images m'ont frappé. Le cosmonaute hors de sa cabine par exemple, flottant autour : sur un mode épuré, bizarrement, il interroge les liens d'appartenance. Il nous suggère de revoir ce qu'on appelle appartenir : la part tenue par lui, la part qu'on tient et qui nous tient, est rarement, à ce point, **élémentaire**. Il est donc là, appendice de sa cabine, fragment terrestre hors de son habitacle, motte de terre pensante qui tourne autour de la terre, et qui est à la fois de la terre et de la non terre. Du reste, elle pense un peu moins qu'elle n'est **pensée**. Disons qu'elle sent l'acuité de ce point zéro, pourtant saturé de science, surchargé de savoir. Ce fragment de planète pensante qui plane dans le vide hors de sa capsule ferait presque vibrer une certaine corde religieuse (dans le genre : poussière tu es et poussière tu redeviens), mais cette fois c'est de la science que viendrait ce petit rappel trivial, cette image mise en acte. Que la science et la technique soient habitées par de l'élémentaire, de l'archaïque, tout chercheur le ressent, et c'est avec cela qu'il **découvre**. Mais ici, la chose mise en acte est collectivement vécue : c'est toute la terre qui voulait toucher la lune (même si certains hommes de tradition, des religieux nullement fanatiques se sont rebiffés lors du premier alunissage : "*Non, ils n'ont pas pu mettre le pied sur la lune ! La lune est une lumière...*", comme s'ils voulaient la sauvegarder, qu'elle puisse rester peuplée de mythes nocturnes : qu'on puisse encore demander la lune...). Soit dit en passant, **atteindre** la lune, la toucher, fut une effraction poétique, une secousse réelle de langage : mais rien de technique, je crois, ne s'en est suivi, nul n'est allé y installer quoi que ce soit. Comme s'il suffisait d'y loger le **geste** de l'avoir atteinte. Autrefois on nommait le lieu nouveau, Amérique, Colombie... du nom du premier voyageur, un nom qui prenait corps autrement. Aujourd'hui nommer ces lieux c'est simplement nommer le geste de les atteindre... en accomplissant ce geste.

J. Schneider and M. Léger-Orine (eds.), Frontiers and Space Conquest, 189–195.
© 1988 by Kluwer Academic Publishers.

Un autre élément m'a touché : le rapport à l'horizon. D'emblée on pense que le cosmonaute vit une absence d'horizon -et l'angoisse qui s'ensuit d'un espace unifié sans **coupure** et sans **autre**. En fait ce qu'il vit est plus subtil : il vit l'horizon fragmenté : dispersion d'horizon, fragmentation. Je m'explique : l'horizon cesse d'être la ligne figée qui fait signe d'un au-delà et qui, pour garder cette valeur de signe, s'éloigne lorsqu'on s'en approche, s'approche quand on s'en éloigne... sur terre. Là on s'en éloigne dans l'**autre** dimension.

Le cosmonaute voit l'horizon, cette coupure essentielle de notre espace, il la voit s'incurver de plus en plus, devenir ronde et du coup il perçoit sûrement la courbe ronde des autres planètes, notamment la lune, il la voit, cette ligne d'horizon, il la voit moins courbe, un peu plus plate parce que plus proche. De ce fait, l'horizon, cette coupure fondatrice de notre espace, cette frontière du rapport à l'Autre qui n'est pas nous et qui ne peut que nous échapper, voilà que dans la levée de l'astronaute elle est restituée à sa fragmentation archaïque : ses fragments sont comme des lettres d'un alphabet naissant.

N'est-ce pas d'ailleurs avec ces bords fragmentés que travaillent non seulement le topologue mais l'architecte ? L'architecture d'un espace est portée par ses bords, ses ouvertures, ses orifices pulsionnels, tout comme un corps jouissnt et souffrant par ses bords cernés. Construire un espace ou le faire advenir c'est faire parler des fragments d'horizon. De l'horizon comme blessure d'où se prélève dans le vide autant d'espace qu'on **désire**. Car à travers son horizon, tout espace est d'abord lieu de rencontre avec l'Autre. L'autre face du monde par exemple, ou des choses. Disons en passant que grâce aux satellites qui le cernent, le monde a comme perdu la face cachée : sa surface est mise à plat. On peut **voir** ce qui se passe de l'autre côté : cela ne donne pas forcément une seconde vue, mais il fallait le faire.

Car un des retours sur nous de cette plongée dans le cosmos, ce cosmos dont le silence connu pour être effrayant s'est mis à être très attirant, comme tout ce qui fait peur et qui nous confronte à notre horizon. -à nous-même comme horizon de nous-même- ce qui nous en revient de plus patent, de cette plongée dans le vide, c'est qu'elle fait retour sur elle-même : le regard lancé vers l'infini se rebrousse vers la terre : jamais la terre n'a été aussi **regardée**, observée, cernée, guettée, qu'avec le dense réseau de satellites qui lui sert de voisinage. La terre se retrouve enchâssée dans ce réseau de satellites qui lui renvoie les regards issus d'elle : occasion comme une autre de **réfléchir** non seulement sur soi mais à travers soi : à travers ce décentrement.

Est-elle mieux gardée pour autant ? Mieux sauvegardée ? En tout cas on l'observe avec l'attendrissement qu'on a pour une belle machine absolument **unique**, totalement **singulière** qui nous appartiendrait autant qu'on lui appartient, même si cette double appartenance se trouve elle-même béante, brisée, décentrée d'elle-même. C'est plus qu'une manière de vivre le fameux écart copernicien. C'est l'expérience radicale où plus l'homme va vers l'Autre, le tout-Autre, plus son regard exorbite revient sur lui. Et ce qu'il regarde avec surprise ou hébétude, c'est son point de départ, auquel il se retrouve **aliéné** c'est-à-dire simplement devenu autre.

C'est qu'à prendre pour miroir l'immensité du vide cosmique on doit pouvoir recueillir, en retour, de curieuses images de soi. Des prises de vue plutôt louches.

Cette étrange filiation entre la terre et le cosmonaute... Déjà au départ de la fusée, la terre le **largue** autant qu'il la **laisse tomber**. Je l'imaginais accouchant de la terre, et celle-ci l'éjectant, l'expulsant loin d'elle. La terre est hors de lui en lui. L'enfant né hors de la mère qui reste en lui -un certain temps semble-t-il.

Le cosmonaute, à la limite, risque de nous apparaître enfant psychotique de la terre. Quelqu'un me demandait à quoi peut ressembler l'inconscient d'un être né en état d'apesanteur ? Comme si la question portait sur l'avenir lointain. Or ce genre de question, dite de science-fiction, porte plutôt sur le passé trop proche ou le présent immédiat. Car enfin, la gravitation terrestre, cette façon qu'a la terre de faire peser les choses, a plus d'un sens, et ne concerne pas seulement la chute des corps (Rappelons tout de même que dans certaines langues, l'enfant qui naît, on dit qu'il tombe, hors de sa mère, comme si la Terre l'attirait dans son champ, hors du champ maternel). En tout cas, naître dans un champ de forces qui vous retienne, cela fait sens. Il y a des êtres qui naissent dans un champ de forces quasi nul, ou rien ne les tient ou ne les retient. Ils naissent au point zéro des champs de forces humaines, terrestres, qui nous poussent et nous rattrapent. Ils naissent en quelque sorte allégés d'eux-mêmes, malades d'absence (Chez d'autres enfants malades, cette absence radicale se lit à même le regard : regard vide, cosmiquement vide, où des adultes viennent lire leur propre détresse). Mais enfin imaginez le cosmonaute dans sa cabine, allégé, absorbé de longs moments pour attraper un objet, entreprenant le long voyage d'amener un stylo vers ses doigts. Il m'évoque des nourrissons apprenant à occuper leur espace, c'est-à-dire à le faire exister. **Prendre** est une des pulsions les plus riches, déjà chez le nourrisson qui pourtant lui donne un but absolu : sa bouche : pour mettre en bouche. Alors que l'adulte, lui, raffine, fignole : il **prend** conscience, il **prend part** à toutes sortes de choses... Oui, dans sa cabine, il m'évoque le nourrisson, ou certains psychotiques. Après tout la psychose est une production humaine, et ce n'est pas de l'enfermer qui fait qu'on en est coupé ou protégé. La psychose est une de nos modalités d'être, même si la plupart ne la fréquente que par la peur, la terreur du noir cosmique ou utérin...(La peur qui tourne parfois à la fureur : **asile** est un beau mot, même si on s'est arrangé pour que sa réalité soit horrible : mais que les fous aillent trouver asile dans des lieux pour se protéger de la folie furieuse des normausés, l'idée n'est sûrement pas folle). Le psychotique, et le néo-psychotique que sera sans doute le spatiopithèque, c'est en un sens quelqu'un qui n'a pas d'inconscient parce qu'il est l'inconscient d'un autre, le fragment d'inconscient d'un autre. De plus, l'humanité surtout moderne a tendance à passer à l'acte ce qu'elle n'assume pas, ses fragments d'inconscient. Les cosmonautes dans leur capsule seront des héros parmi d'autres de ce **passage** mais encore une fois, étrange effet de retour, leur regard, le regard qu'ils sont, se retourne vers la terre et la **fixe**, lui donne à voir dans un miroir cosmique sa peur d'avoir un inconscient, sa manie de s'en débarrasser, au point de devenir soi-même l'inconscient de quelqu'un d'autre ou pire, l'inconscient de personne.

Notre expérience de l'horizon, ordinaire, est celle d'une simple limite : et l'expérience cosmique nous montre l'horizon comme pluralité de limites, fragments hétérogènes. Cela confirme que cet élan spatial n'est pas tant l'élan compulsif pour reporter la limite, "reculer" les bornes de ce que l'homme peut atteindre. Il y a là plus qu'une suite d'inclusions dont chacune réfute la précédente et attend d'être refutée par la suivante. En fait, l'homme semble **chercher ses limites**, comme s'il ne prenait pas au sérieux celles qu'il a déjà : comme s'il avait peur de les voir craquer, s'estomper : comme s'il feignait de le croire. Alors il en cherche d'autres plus solides. L'intérêt d'une limite **autre** c'est d'offrir un appui **autre** ; un appui différent pour faire retour différemment, à soi, sans que ce soit un **retour au même**.

Dans cette aventure, cette "conquête", l'idée n'est pas d'étendre son espace : la terre ne **couvre** rien sinon elle-même dans cette lancée. L'idée forte n'est pas celle de l'extension, mais du jet, du trajet, des flèches, des orbites, des percées, des pénétrations ponctuelles où l'enjeu est d'aller à la rencontre d'un **point-limite qui soit de retour** : d'un point de rebond fécond. Aller à la rencontre d'une loi qui tienne, d'une limite qui tienne assez pour être un point de rencontre avec l'Autre. Ici l'Autre c'est l'autre-dimension, pas facile à qualifier, peut-être inqualifiable. Ce n'est pas vraiment la "verticale" : les avions déjà prennent la verticale partielle, avant de filer à plat ventre comme tout le monde, à l'horizontale surélevée. On aimerait l'appeler dimension temporelle, mais elle est aussi spatiale. Disons que c'est la dimension autre, l'autre dimension, à travers laquelle il y a peut-être à conjurer, à entamer ce que l'Autre comme tel a d'absolu (Certains pensaient ouvertement, en sillonnant le ciel, y entamer la toute-présence divine...).

Toute percée vers une dimension autre, fait apparaître l'espace de départ comme une "surface". Et à chaque saut, la surface terrestre se trouve, en tant que face, toujours plus surdéterminée : fibrée sur un mode plus complexe.

Autre idée simple et brutale que l'astrophysique souligne : depuis qu'il existe, l'homme est mis devant des tâches qu'il passe sa vie à n'accomplir qu'en partie, histoire de passer la main, de transmettre même à son insu aux générations suivantes. Mais face au voyage cosmique, il est devant une ligne, une trajectoire toute tracée, d'ici à telle planète qui exige **plus d'une vie** pour être simplement parcourue. La simple orbite met en ligne plus d'une vie, des générations, des temps **excessifs**, des excès de notre espace sur lui-même. Heureusement qu'il y a pour surmonter cela le vaisseau increvable de l'espace psychique, avec rêves, fantasmes, folies, pensées... projections ajustées, confrontations au grand vide où se prélèvent tous nos espaces, et où fut prélevé il y a des millions d'années ce petit monticule sur lequel on s'agite.

Si notre espace s'excède, nos lois aussi. Les lois de la nature et du cosmos sont comme des jets différentiels **infinis**. Ce qu'on nomme ici, d'ordinaire, ce sont les termes finis qu'on parvient à écrire. On échoue à les écrire tous. Leur totalité, si même elle est pensable, excède nos moyens d'écriture : mais nous incite à les renouveler. Parfois, telle tradition mythique feint de formuler des lois ultimes, mais personne ne s'y trompe : les lois mythiques ne sont pas fausses, simplement elles ont enfermé en elles, peut-être pour le préserver, le point d'infini. Elles en tiennent lieu, et laissent croire ainsi qu'il est formulable, qu'il est en elles formulé. C'est une façon de tourner la question, mais cela n'empêche personne de la retourner, ladite question, celle où se lient **amour** et **connaissance**.

Qu'est-ce qui a précipité l'humain dans le vide cosmique ? (Ici ma machine fait un lapsus : dans le vide **comique**...). Une idée simple m'a séduit : le gain de vitesse des vecteurs humains a soudain débordé la terre. L'espace terrestre fut comme gagné de vitesse, dépassé par ces champs de vecteurs nouveaux, de plus en plus accélérés, dont la vitesse ne fit plus sens à l'échelle terrestre (à quoi rime d'aller de New-York à Tokyo en trois quarts d'heure pour 5000 dollars ? D'autant que le mouvement des corps réels fut aussi pris de vitesse par la transmission des lettres, des voies : télé, télex...). Donc l'impulsion des vecteurs humains a débordé la ronde terre, et l'espace cosmique s'est présenté comme seul **autre** possible, pour dialoguer avec ces vecteurs. Effet surprenant du dialogue, je l'ai dit : on a peu vu des planètes, mais la terre elle se fait voir sous tous les angles. La Terre ceinturée de satellites, la Terre ceinte de son regard sur elle-même, de son écoute d'elle-même, de sa tension exorbitée, mise en orbite de satellites qui la palpent de toute part.

La Terre a satellisé ses produits de pointe : ses appareils les plus futés et affûtés pour l'ausculter, pour simplement la mettre à plat : elle a surmonté sa rotondité, totalement. Déjà le télégramme avait fait des siennes : une de mes amies reçut un télégramme de Nouvelle-Zélande : "Père mort à 9 h". Avec le décalage horaire, elle le reçut à midi et dut attendre sept heures durant, que son père meure à son heure, à elle.

A propos de cette conquête, certains ont parlé de Tour de Babel. Est-ce bien l'image qui convient ? Bien sûr le fantasme de langue unique, monolithique, est à l'oeuvre : technologie, ordinateur, totale programmation... Mais l'immensité du but, son absence presque, le ciel trop vide pour être atteint, tout cela et d'autres choses comportent un pouvoir **dispersant**, qui couvre d'avance de ridicule toute prétention babélienne. Quelques humains s'éloignent dans leur suppositoire. On prie pour **qu'ils reviennent**. Leur retour est déjà le comble du **challenge** : qu'ils reviennent après **ce saut qualitatif** et qu'ils puissent, eux ou leurs outils, nous en dire quelque chose. Car l'important dans une perte de contact, c'est moins la perte que **le devenir langage** de cette perte. Y a-t-il dans ce devenir quelque chose qui ne soit pas réductible à la pure technologie ? Au pur langage programmé ? Au pur travail opératoire ? Y a-t-**il** ? Question ouverte.

Un signe déjà est évident : le désir de parler et de penser **cela**. C'est un cas particulier d'un désir plus vaste qui vibre dans toutes les sciences : le désir de philosopher, de penser ce que l'on fait. C'est au point que par un effet de retour, l'acte de philosopher retrouve ce qu'il a d'**élémentaire**, de radical : non seulement redonner sens, ou assister à l'engendrement du sens, mais savoir en être partie prenante, du renouvellement de sens, à travers l'expérience insensée.

Aujourd'hui c'est du fond de l'acte scientifique et technique qu'émerge un désir de penser l'acte et ses après-coups. Ce désir de philosopher (c'est-à-dire littéralement : de se ressentir **aimant** penser ce que l'on fait) est vaste : appel à se confronter avec d'autres modes de penser, d'autres expériences de penser : mais aussi peut-être : désir de faire son analyse au moyen de la recherche ou de la pratique que l'on mène. Prendre une part de soi pour s'analyser d'autre part... On est loin d'une épistémologie qui court après la science pour lui dire ce qu'elle fait avec cette fausse honte de l'ignorant qui ne sait rien faire lui-même.

De quoi en tout cas torpiller les concepts enkystés, les bavardages gris qui n'appellent qu'à eux-mêmes. Or qu'appelle-t-on penser sinon la rencontre d'une question vive qu'on lance et de questions plurielles et plus vastes qui étaient là en attente, en orbite d'attente ? N'est-ce pas l'appel ou le rappel qui surgit de cette rencontre ? En ce sens, la **conquête spatiale** peut être d'abord la conquête d'un espace psychique renouvelé. Jusque là, on est loin des grands émois, des tremblements de curiosité et d'angoisse d'un Colomb et des siens se demandant ce qui les attend à l'**autre** bout de l'Océan. L'ennui, mais aussi le singulier avec la conquête cosmique, c'est qu'on n'est attendu nulle part : c'est que l'enjeu est de pouvoir, de toute part, d'où que l'on soit, garder le contact avec l'autre part de soi-même, avec la Terre restituée à son énigme originelle.

Contrairement à Colomb et aux conquérants du Nouveau Monde, on n'est semble-t-il attendu par personne d'autre. De sorte qu'il n'y a pas d'**autre** à exterminer (pas d'Indiens). Il y a plutôt l'expérience de l'autre à faire, à préserver, à refaire autrement : à inventer. Or faire l'expérience (au sens de la vivre) est aussi compliqué que de supporter un choc. Du reste, expérience et trauma sont deux pôles de tension, toujours actifs dans la texture de nos vies : et leur lien est paradoxal : vivre réellement une expérience est presque impossible. C'est passer du côté **trauma**, là où on est débordé et où il faut par la pensée saisir le bord : c'est presque impossible car il faut comme assister à l'instant de notre absence pour pouvoir en témoigner... ne serait-ce que pour nous-même. Vivre une expérience, c'est refaire celle de l'origine. Et peut-être la cosmique met-elle en acte une renaissance : ou une naissance à des temps aigus de nos vies, à des retours au monde, des retours à soi-même autre, bref à une manière de redécouvrir ce que l'on a déjà (n'est-ce pas le don le plus difficile à faire, que de donner à quelqu'un ce qu'il a et qu'il ne sait pas avoir). Par le cosmonaute, la Terre se redonne à elle-même, en croyant se donner au ciel. Grâce aux voyages cosmiques, nous **savons** -nous pouvons savoir- que nous avons la terre, et elle nous est restituée par des voies singulières. Vivre, en un sens, c'est ne pas cesser d'en revenir, de ces découvertes-là, de ces mises à nu de l'espace. Cela nous implique comme constructeurs d'espaces, **metteurs en espace** de nos vies car notre ignorance de l'espace est elle-même archaïque. Même la mathématique nous en dit peu sur l'espace. Elle sait nous en dire sur la mise en espace de tel geste, de telle situation, de tel symptôme. Mais tout le reste est béant, indéterminé. Notamment comment on devient l'espace de sa vie : et le rythme (c'est-à-dire la pulsation de cet espace) en forme de temps et d'histoire. L'espace donc s'engendre à partir des vides immenses, originels, par des prélèvements d'horizon, des cernages, des contours, des bords débordés, assez articulés pourtant pour capter des lieux, en faire des lieux de rencontre.

Et à travers le vide cosmique, il se trouve que l'homme fonce à la rencontre de ses limites, donc de lui-même et de ses textures faites avec ces limites. Il fait l'expérience de ceci que l'engendrement du lieu et celui du langage sont de même nature.

Il se trouve que certains font ce voyage orbital, planétaire vers leurs limites qu'ils capturent, sans bouger : sur place. Les grands toxicos et leur "planète". Dans un livre récent, j'ai développé l'idée que ce qu'ils cherchent c'est un lien et un lieu dont ils soient l'unique auteur. C'est comme s'ils voulaient une petite planète pour eux tout seul, où planer comme des petits princes, une planète où le temps serait l'instant. Il est comique que dans la langue des toxicos, on nomme **capsule** ce qu'on prend pour faire le "voyage". Et l'expression "s'envoyer en l'air" n'est pas si naïve pour exprimer le dégagement. Mais là, c'est l'humanité qui se drogue avec ses capsules, hautement technologiques, avec lesquelles elle s'offre une hallucination **vraie**, et dont elle recueille l'étrange questionnement sur elle-même, et sur ce qu'elle peut supporter comme Autre. Question sur son pouvoir de se supporter comme Autre.

Elle en recueille la jouissance d'aller inscrire là-bas, très loin, les traces d'une loi qui vient d'elle. Jouissance de faire aller loin ce qui nous est le plus proche, le produit familier de nos mains et de nos pensées.

Et c'est une expérience toujours neuve, celle de toucher au réel, qu'on croit avoir surmontée, par des calculs, de la pensée opératoire... Or dès qu'on y touche, dès qu'on met le doigt dessus, c'est le choc, on reçoit une décharge d'**évidence**.

A PERSONAL ACCOUNT OF SPACEFLIGHT

Jeffrey A. Hoffman
NASA Astronaut
Lindon B. Johnson Space Communication Center
Houston, Texas 77058, USA

I will first make a formal personal introduction. I am Jeff Hoffman. I am a NASA astronaut. I joined the Astronaut Program in 1978. I first flew in April of 1985. Because of some of the ramifications, changes, uncertainties that occured in our Space Program, I actually trained for four different flights, all of which were either changed or cancelled. I also was assigned to the very next flight after the Challenger explosion, which would have been an astronomical exploration of Halley's comet and other objects. And I am now, of course, waiting to carry out this flight at sometime in the future. So when people ask me what is the most important psychological characteristic for an astronaut, a space explorer, I would have to say it is patience. Hopefully this will not always be true.

I would also like to make a philosophical introduction because of the nature of this symposium. When I think about my own philosophical ideas, and philosophical roots especially as pertaining to the Space Program, I realize that a great deal of them actually come from the reading of science fiction. I am glad Ben Finney did pay some attention to this in some of the slides he showed. I would suggest that this is a fascinating week we have spent here, but the one thing that has perhaps been missing if we talk about the impact of space on the human mind, the human imagination, is the fact that people have been considering these questions already for generations in the form of science fiction and that we really have to look to this source as well as to academic formal philosophy to see how people perceive the varieties of human experience in the environment of space. The one other thing I would say about myself by way of introduction, is that I have wanted to be an astronaut ever since I was about 6 years old. In contrast to a lot of my colleagues, who actually started their careers as pilots, I was never particularly interested in airplanes. I was interested in rocket-ships. In this sense, I find that I am different from a lot of my pilot astronaut colleagues. I will mention, however, that when you look at the new generation of people who are applying to be astronauts, you find more and more people who have dreamed of space travel for themselves since their childhoods. And this perhaps will cause some changes in the population of astronauts in the years to come.

J. Schneider and M. Léger-Orine (eds.), Frontiers and Space Conquest, 197–208.
© 1988 by Kluwer Academic Publishers.

Why am I here? What do I have to share with you? Obviously, I had experiences that very few people have had, and people are interested in these experiences, and I want to share them with you. Taking it by and large, I would say that the experiences that I have had are things that which I was able to imagine intellectually before I went into space. Sometimes it is hard to describe an experience. You have to realize that what I underwent in most (but not all) cases, was to learn what an experience felt like, whereas beforehand I had only been able to think about and imagine it. It starts from the very beginning. We have all seen launches. You feel the tremendous power even when you are on the ground when the soundwaves hit you and shake the ground and shake your body. But the experience of actually riding the fire of feeling the acceleration of going into space is absolutely overwhelming. Again, it is something you experience instead of just think about. There are also the beauties of space, and we try to share these. We take as many pictures as we can. There is a whole aesthetic experience which I think is fundamentally new, and I think it is important to try to share. I hope that someday these experiences will become the provence of artists as well as just astronauts taking pictures. And then of course, there is the experienced of weightlessness, which is fundamentally new, and which, as much as you try, you cannot imagine. It is almost impossible to communicate. I think the expression Professor Seidengart gave us of weightlessness being a "*state of grace*" came the closest to any way I have heard it expressed in any language. It is something which is absolutely extraordinary. When I think of all the things back on Earth I miss about space, it is not somehow the speed, its not the views, these things I can somehow bring back psychologically. But I have never been able to bring back, even in my dreams, the exquisite feeling of freedom that you get when you float weightless. I was fortunate enough to go beyond this experience inside the shuttle and be able to put on a spacesuit in what I consider for myself personally, the most intimate experience of the space environment that I have ever had, on a 3-hour spacewalk. I will have a little more to say about that later.

There is another aspect of the space experience of being an astronaut which maybe you do not think about a lot, but I find that it is very important. That is trying to communicate the experience. I have not kept track of how many people I have actually talked to since my spaceflight. It is in the many thousands when I talk to audiences, and when I count radio and television audiences it must be in the millions. I think this is a very important part of the experience of being an astronaut. It is a very important part of the entire Space Program. The reason is that NASA does not send me into space in order to have a philosophical or meditative experience; they send me there to do a job. But I feel that part of what I am doing is also to try to share with people who do not go into space something about the experience. If our experience does not somehow enter into the wider human psyche to become part of the experience of the mind of humanity as a whole, then the importance of what we are doing is lost. To use the analogy of the exploration of the New World, historians love to argue about whether some Scandinavian Vikings or perhaps some Irish Monks in the 12th century may have actually landed in the New World. But in the important historical sense, whether they did or did not really does not matter, because it never became part of the European mind. People did not know about it; it did not make a long-term difference. I think if what we are doing in space is to make a long-term difference then we have to share the experience. It has to become part of the human mind. Part of the experience I have had in sharing this is to see the tremendous enthusiasm, the interest people have in the experience of space. It is extraordinary to watch people, especially young people, waiting in line sometimes for an hour to get the chance to ask questions about the experience. What is it like? What can I do to share this experience? People have the most extraordinary ideas of space. Sometimes we transfer our psychological desires into space and sometimes make the mistake of seeing space somehow as a way to get away from human problems of the Earth. But people have these ideas. Space plays an important role in a lot of people's psychology and I see this directly. I am going to talk about several metaphors today, and one of the metaphors which has occurred to me partly through this experience of trying to share my experience as being an astronaut, is the idea of the space ship as the Gothic Cathedral of our age, as somehow being the embodiment of the psychic force of our 20th century technological civilization. There was a reference by Ben Finney to the fact that C.S. Lewis and some other people might disapprove of, the Space Program. Lewis certainly disapproved of the 20th century as a whole and he looked back to the time of the Gothic Cathedrals and the Medieval Europe as the last time when humanity somehow, at least western humanity, possessed a unity. Philosophers talk about the myth of the lost unity of the Middle Ages. Well, if there is anything we do today that unites humanity, in terms of dreams of the infinite, something all people can do together, it is our exploration of space. I think we have seen this at this meeting this week. We see these space ships. They are the embodiment of all that is finest, of all that is most perfect about our technological civilization. Now when I say this, of course, it is no accident that this and the last picture I showed were of the Space Shuttle _Challenger_. We cannot think of the Space Program without the knowledge that space ships do not always work. The Russians and we Americans have suffered tragedies. I think that we have to remember, continuing my metaphor, that Gothic Cathedrals sometimes fell down. The architects learned by a process of trial-and-error and when we look back, what we see is not the Cathedrals which collapsed, we remember and we still appreciate the beauties of Notre Dame of Chartres, and I hope that when we look back on this stage of the exploration of space that what we remember are the beauties and the successes and not the disasters.

I will digress a little bit from my philosophical comments now and personal impressions, because I think it is important after mentioning the _Challenger_ tragedy to look ahead briefly and talk about where we are going once we resume space flights. Of course, the Shuttle has launched a lot of satellites in space. We have a big backlog of satellites which still have to be launched. We will now be sharing the job of launching these satellites with expendable launch vehicles. But still, at least for the first few years of Shuttle operations, a large part of what we do will still be to take satellites up into space. I should say on a philosophical bent when we talk about the impact on the human mind on what we are doing with the Shuttle that probably everybody who thinks about it knows about the importance of satellites, but how many people have actually seen a satellite in space? I think that these early pictures of satellite launches from the Shuttle do date the first time anybody had really seen a satellite in space. I think that is important when you talk about people's perception of space and what we are actually doing. But I hope that if we have learned anything from these experiences, it is that, for the present at least, with our current level of technology putting human beings into space is an expensive and still a dangerous enterprise and we would be ill-advised to do things with people that can be done without people. I hope that the emphasis of the manned Space Program will become more and more on those things for which you need people in space. I am referring partly to scientific activities such as those carried out in the Spacelab. I hope we have many more of these Spacelab flights. Many astronomical observations depend on satellites. In this area I think the Shuttle will continue to be very important. The idea that we can rendezvous with satellites, repair them, refuel them, and change out their scientific instruments, brings a whole new dimension to human possibilities in space. I look forward to being able to use these possibilities more fully in the future. The other thing which is so important when we look at the future of space technology is the ability to use a spacecraft like the Shuttle, or perhaps Hermes in the future, to test new space technology: to actually try out things and see how well they work and adapt them to fly over and over again if necessary until they work; to test out new techniques, things which we just cannot do on Earth because of the presence of gravity, such as new techniques of constructing structures which cannot even support their own weight. This is something that is fundamentally new, something that can only be done in space. Of course, the near-term culmination of a lot of these technological experiments will be the Space Station, which will be a whole new jumping off point. Of course, it is something that the United States had back in the early 1970's. We unfortunately lost the capability to operate for a long time in space and I look forward to the time when we get it back. I admire, and have to admit to being somewhat envious of our Soviet colleagues, who for a long time have maintained and developed this capability. I have great admiration for what they are doing and I look forward to the time when the Western World can join them in this endeavor.

Let me go back now to some of my own experiences of spaceflight, to my perceptions, the things that I felt were important to me. Spaceflight certainly changes your perception of time and space through the great speed of orbital travel. Now obviously the biggest change philosophically and scientifically in our ideas of time and space in this century have been the theories of relativity. The speeds of current space travel are very small compared to that of light. Relativity may some day become important when we talk about interstellar travel, but right now the relativity I talk about is not scientific relativity, it is psychological. Our perception of time and space when we are going around the world every 90 minutes is obviously changed. I do not feel this is a revolutionary psychological sense. I think that speed in the modern era has been revolutionary but I think this change has really already occurred and has occurred before the Space Age. The fact that I can come from Houston, spend a few days here in Paris at a colloquium and then go back to work next week, that is fundamentally revolutionary, more so than the fact that I can go around the world every 90 minutes. But nevertheless, it is a unique experience and it cannot help but gives you the psychological perception of how samll the Earth really is. This is something that you know intellectually before you go, but the actual experience of it is so much stronger than the idea you had beforehand. But there was something else that was strange that happened to me: despite my feeling of how small the Earth was, I was constantly amazed at how big the oceans were. You go around the Earth every 90 minutes, and it seems like a short time, and the Earth seems very small, but the fact that it took 30 minutes to go across the Pacific ocean seemed to me like an eternity. And you really do realize that the Earth is the Blue Planet. The oceans are big. The oceans are also important to me because so often when we think about space travel we use metaphors of ocean travel, from the Polynesian explorers, the exploration of the New World, and also going even further back in European mythology, the voyages of Ulysses, and Jason's Argonauts. The oceans have always been the symbol for the unknown. Beyond the ocean is where we find adventure. New ideas could be located on the island of Atlantis. The monsters of the imagination could live on the other side of the ocean. When we think about how future generations will look back towards space travel, I think it is important to recognize now that this age of the Sea is over. We no longer look at the oceans in this way. And I do have to wonder what will be the future of space travel. Will we come up against some limits so that we will end up staying in the inner solar system? In this case, probably sooner than we are willing to think, space also will cease to be a source of human imagination and we will come to accept the new environment that we have mastered in the same way we now accept the fact that we have a mastery over all the Earth. I hope of course, that this is not what will happen. I hope that the image of space as infinite will some how come to fruition and that it will somehow turn out to offer mankind much greater possibilities, something fundamentally new, in the human experience and in the human mind. But I do not know. At any rate it is something fun to dream about. Once again, it is to science fiction that we have to turn for well-developed ideas of where we may really be going if some of these things happen.

I would like to talk about something else I experienced in space which to me was absolutely overwhelming, my impressions of the atmosphere. When you look at this picture of the Earth, what you see most of all is the vast expanse of the ocean. But look up at the horizon now and if you look closely you see between the bluish-white of the haze on the edge of the ocean and the black of space, a very, very thin blue line. I was amazed when I was riding into space in the rocket and looking out the window still within the first two minutes of flight while the solid rocket boosters were still firing, we were at an altitude of less than 30 kilometers and already the blue sky of Earth had turned into the black of space. Often we talk of the Earth as being a cocoon for mankind, but when you look at it from space, you see that the skin of that cocoon is very thin indeed. I never ceased to be overwhelmed by the experience of how small the sky looked from space. I do not know what sort of metaphor you have in French, but in English whenever we want to talk about how big something is, its always "as big as the sky". When we talk about the sky, we are talking about the blue sky, the atmosphere. But the sky is not big. The sky is very small. And although you can see this in pictures, somehow the experience of actually seeing the sky beneath me, I found absolutely overwhelming. I never got tired of looking at the sky and all of its manifestations, the sunrises and sunsets which we experienced twice every orbit. The beauty never failed to delight me.

I am an astronomer, and a lot of people ask me, "What does the sky in space look like at night"? This is the best image I have to share with you. Of course, the picture was actually taken scientifically to study the aurora which is quite spectacular, but I ignore that for the moment and point out to you that when you look at the Earth at night, and you look at the horizon, by and large, this is what you see. You have to look very closely to notice that beneath the "horizon" you can see stars. This is not the horizon at all, although this is what catches your eye. The true horizon of the Earth lies down here. What we actually see is an airglow layer about 60 kilometers above the surface. It is a halo that surrounds the entire Earth and you only see it at night. It starts about 8 minutes after the sunsets and it is dark enough so that you can actually see this. For me, psychologically, the most direct experience of feeling the Earth as a planet located in space was at night, when I was no longer aware of the sun, which otherwise dominates the whole environment. To be able to see, first of all, the lights of the stars above and then the lights of civilization below (and unfortunately we just do not have good pictures to illustrate the magnificent beauties of human civilization at night where you have a filigree of lights spread out like gossamer spider webs all over the surface of civilized parts of the Earth) and then between the sky above and the Earth below you have this almost etherial orange halo surrounding it all... it is almost an otherworldly experience that I can best share with this picture. When we look at the atmosphere and realize how thin it is and also how fragile therefore it is, I think it is even more important to continue our exploration of the other Earth-like planets of the solar system.

To understand what it is that made the Earth like it is instead of like Mars or Venus. The importance of understanding the Earth as a planet becomes even clearer when you look at the Earth from space. And here again I get into something which everybody knows intellectually, but the experience of seeing it is an overwhelming emotional experience. What I am referring to is the fact that when we consider the condition of the Earth as a planet and we look at not only the past evolution of the Earth but look at the future evolution of what is happening to the Earth's environment, we have to realize that humanity is no longer just a small perturbing force on the face of the Earth. I could spend a whole hour just showing you pictures of the environmental impacts that humanity has had on the Earth. I am going to confine the images to the clearing of the jungle. Again, we all know that it is happening. Intellectually, I knew it before I went here. I had even seen pictures of it. But somehow to actually fly over the Amazon and see the jungle being cleared was a very frightening experience. I shall share with you just a few of these images to give you an idea of how it starts and how it is progressing. The clearing starts along certain passage ways which are cleared through the jungle, either river ways or, in many cases, highways which are built into the jungle. The growth of the cultivated areas is almost organic, it precedes along little dendrites that grow out perpendicularly from the initial development, and then you get dendrites on the dendrites and ever-increasing perpendiculars spreading out further and further, and then the cutting of the trees procedes until the amount of forest between these dendrites gradually decreases to the point where finally you have no forest left at all. This is something which we see all over the world. It is happening in the Amazon. It is happening in the Congo. It is happening in Southeast Asia. The forests of the Earth are being cleared, they are dying every place you look. There is burning. You see the fires. You see them during the day. You see them during the night. Astronauts who flew back in the 1960's and who have flown in the 1980's have said there is no question in their mind that the view of the Earth, the atmosphere, has fundamentally changed. It is much hazier. You cannot see, even from space, with the clarity with which we were once able to see. It would be naive to suppose that all we have to do is to take pictures of the Earth from space and that this will promote an ecological consciousness so that all these problems will be solved. I do not know what the answers are. Clearly however, without the consciousness that these problems exist there is no way we can solve the problems. I get depressed when I look at these images for too long, because the magnitude of what is happening is gigantic. Clearly if we do not solve these problems, all the excitement, all that is important to the imagination of what to do in the future in space is not going to happen; because we would become wired in the ecological disaster that we will have created for ourselves on the surface of the Earth. And all I can do is to hope, first that future generations may have some better ideas than we have now and, second, that perhaps the force of life in Gaïa is stronger than we give it credit for and that maybe somehow we will be rescued from our own mistakes. So I leave this depressing though and point out, maybe almost as a hope, that there are things about the Earth which we still do not understand, and maybe there will be a way out of these problems. This is my symbol for it: a rather a unique atmospheric phenomenon which I have not heard a good explanation for, sort of a sideward horizontal tornado shaped in the form of a question mark. I think the symbol speaks for itself.

I shall finish up my views of the Earth by sharing with you my favorite view of the Earth. It is the Sierra Madre Oriental of Mexico. If you look down in the lower right and corner you will see why its my favorite experience, because that is where I was on my spacewalk when I saw this view of the Earth. I talked with you about the fact that we are not sent into space to have the opportunity to meditate or to have philosophical experiences, but I want to share with you a little bit about what happened to me on my spacewalk. I turned out to be very fortunate. We were sent out in an attempt to rescue a satellite. We had to attach some objects which we called "flyswatters" to the end of the Shuttle's manipulator arm. I will not go into any details, but the point is, it took us approximately 1 hour and a half to accomplish that task. When we finished, we were just leaving the daytime period and entering night, and the controllers on the ground wanted the opportunity to look at our handywork and to see how it was done. They also felt it was necessary to berth the manipulator arm in order to make sure there was no interference from these tools which we had attached to the end of it which would stop the Shuttle's doors from closing at the end of our flight. They wanted to look at it during the daytime, and that meant that we had to wait, as it turned out, another whole orbit. All of this activity was going on way overon the other side of the Shuttle, which was out of this picture, and my colleague on this spacewalk went over to assist with those activities, so I was in the rather unique position which, in all honesty, very rarely happens on a spacewalk, of having absolutely nothing to do for over an hour except just hang on and watch the world go by. I did a few other things. I climbed up and down the tail and went around and looked at the engines and over at the wings, but it was absolutely an experience which I had never expected to have, which I will never forget. And the impression I carried away with it most strongly was in a sense, a very strange feeling of relativity about what I was; where I was; and what I was doing. And I will explain it this way. If I was holding on to the Shuttle and moving my arm to try and make myself move around, then the mass of the Shuttle was so much larger than my own mass that essentially the Shuttle became the fixed point in my universe and I was just a human being holding on to the Shuttle. But if I let go and just floated or even held on lightly, exerting no force, then I was an independent satellite, orbiting alongside the Shuttle. This was certainly my most direct feeling of being a part of space, of something fundamentally other than the Earth. My strongest impression at that point was of the tremendous importance of the sun. Professor Van de Hulst talked about the need for psychological grounding.

I felt at that time that I was more aware of the sun that of anything else. It was certainly the rythm of the sun which motivated my sense of time, and also the feeling of my body. It may sound simple to say it, but when I was in the sun, I was hot, and when I was not in the sun, I was cold. But that is overwhelming. I came far more than I ever did before to feel, in a very personal sense, the very overwhelming power of the sun. Now obviously we all know the importance of the sun, that the sun is the life force for almost all life on Earth. At his point, I will digress just a little bit and allow myself a philosophical interlude because of the nature of this discussion. What about interstellar migration and the future of mankind? What do we really need if we are to go? Can we ever leave our life-giving sun and go somewhere else? There are two things that I have come to feel about this. First of all, the sun is so powerful, so overwhelming that we will need a very high level of technology to provide a source of energy which will substitute for the sun while we are undergoing an interstellar journey. But we know in principle it is possible, because we have discovered communities deep underneath the ocean which do not require the sun as their ultimate source of energy. So I think clearly it is possible for organisms that ultimately started out as Earth-life based in the sun to harness other energy sources and use them to exist in habitats fundamentally different. What about the other requirements? Here maybe I am drawing on my own experiences as a space traveller. I do not think now we are psychologically ready for this journey. No matter how wonderful the experience was of being in space, whatever I say about the "state of grace" of weightlessness, nevertheless I wanted to come home. In all early voyages even in the early myths of Jason and the Argonauts and Ulysses, they all wanted to come home. But it also was true that ultimately we got to the point in human history where, for whatever reason, people became willing to make journeys and not want to come home. I feel spiritually that some day we wil get to the point of being able to do this, but it will not be travellers like myself who left the surface of the Earth, because I think we are tied to gravity. Ultimately it has to come from people who have spent their lives without gravity. Who somehow have made a decision not to return to gravitational point sources as habitats of life, but to live without this. Once we get to that point, then perhaps we would be psychologically ready to leave the sun and make this journey. Right now, if I asked how many people would be ready to make this journey to the stars, some people might raise their hands. But I think, given the current state of our psychology and our relationship with gravity and the planet Earth, that anybody now who would say that they are ready to make this journey is psychologically unfit to make the journey. I think it would be dangerous for them to attempt it. I certainly would never go with them.

So much for the philosophical intervention. I want to now come back to Earth. In finishing up my presentation, I think there are one or two more things that are important. I will share with you an extraordinary experience that I had the last night before we came back to

Earth. I was floating, looking out the window down towards the Earth at night. And I saw a meteor go underneath me. Now, I am an astronomer, and I have seen a lot of meteors, so I recognized that it was a meteor, there was no question. But then a moment later, I said to myself, "Wait a minute, that can't be a meteor because that was below me and meteors are up there" And then of course, a fraction of a second later this realization hit me, "wait a minute, I'm in space, and meteors you see when they burn up in the atmosphere, which is below me!" I think more than anything else this realization really put me in my position of where I actually was. But it was also a very strange feeling, because seeing that meteor burn up in the atmosphere, I thought about what it was that we were going to be doing the very next day.

Let me now comment a picture taken over the skies of Houston in the early morning. The Space Shuttle Discovery is making its passage, going somewhere around Mach 10, ten times the speed of sound, on its way to a landing at the Kennedy Space Center in Florida. It is experiencing temperatures in the order of 1500°C on the outside and is glowing like a meteor. Now obviously the Shuttle had done it before, and I had every confidence that we were going to be successful in our own reentry, but nevertheless it was a rather unique experience to see that meteor go underneath and realize that the very next day we were going to be inside a meteor. I share just two pictures with you of what it looks like to be inside that meteor. If you look out the front windows, there is the commander to the left. What you see is a red glow. Actually it starts a dull red gradually heats up to a bright red, orange and eventually turns into an almost white light on the outside. This is very diffuse, it is the plasma sheath, the ionized gas in front of the Shuttle. If you enter the daytime, this fades away very quickly. But for me what was even more spectacular was what is visible behind the Shuttle, the wake, the essential the confluence of shock waves that follows us down from orbit. There has never been a really good picture of it because the exposures are so difficult. These are streamers, they are constantly flickering, but the form itself stays very constant. It is almost like a statue with a body and a head, and then right here, where almost the neck or perhaps where the heart would be is a very intense almost diamond-like white light. For me, it was almost a totemic figure. I was absolutely overwhelmed by it. I was standing there almost transfixed looking at it. I should mention that I was not responsible for flying the Shuttle during the rentry, so I had the luxuary of being able to experience this sort of thing. I would have been very upset if I had seen our commander or pilot turn around and look at the same thing. But I called one of my other crewmembers over and asked her to look at it, because she was concentrating on looking out the front window, which was also quite fine. It was almost that I somehow wanted somebody to confirm what I was seeing. She floated back, because at the time we were still floating, and she looked at it, and our eyes were wide open, and she looked at me and said "Do you think we should bow down and pray to it?" That is for me how strong an impression it really made. Well, at any rate, we did obviously survive the reentry and not much later we were landed in Florida. I can never fail to remind not just other people, but also myself, that when a Shuttle lands, you have to realize that a half an hour earlier it was in space. To me that is still an extraordinary thing, an extraordinary accomplishment.

Despite the problems we have had, I think we have come a long way. We have done some extraordinary things in space and look forward to continuing. We will fly again. What I hope is that we have learned enough from what we have done to realize the enormous amounts of power that we have to harness to get into space. At our current level of technology, we are still pushing the limits of the possible. It is not something we can undertake lightly. We have to realize the environment in which are working. I think Richard Feirman in his appendix to the report on the Challenger Accident said, and I think it is very important that we must never forget it in this enterprise of spaceflight. "*We can do anything we want. We can say anything we want to ourselves, because it is easy to fool ourselves. But, we cannot fool nature. And if we try to fool nature, we only court disaster*".

Let me finish up now by going back to our visions of the infinite, the impossible. I talked to a couple of my crewmembers before my flight about how can we actually illustrate that. Did they have any dream of some personal thing they would like to do to illustrate the impossible things we can do in space? I will just show you two pictures, somewhat humorous, as these things tend to be. But there is a deeper point behind it. One of my crewmembers is an amateur juggler. Often when things would get a little bit boring during the training he would take little pieces of flight hardware (on the ground) and juggle them. He is rather good. Well, juggling has a long history, thousands of years. The dream of every juggler since the dawn of mankind is to be able to do it well enough with enough objects that you can somehow make these objects appear suspended in space. Nobody has ever done it until him. This picture, by the way, appeared on the cover of Juggler's International magazine, so it has gotten quite a wide coverage. For me, I have a small collection of oriental rugs at home. I would have loved to have been able to bring one of them along with me; it would have been aesthetically much more pleasing than what I was able to put together out of the Shuttle's equipment. But I think the point is there anyway. How can I sum things up? Somehow you have to go back to art and literature. When I think of the expressions of the impossible, the second-to-the-last-line of Goethe's Faust to mind, "*Das unbeschreibliche, hier wird es getan*" [That which you do not even have words for, here we can do it]. And it is really true! It is a world without limits. This is the dream we have. That is why it is so important to us. That is why I enjoy the experience of being able to share it with you.

In closing now, there is one more part of the experience which I think does have to be shared. I do not usually talk about it. It is a very personal experience, but it may be appropriate here. And it again has to do with reentry. I showed you the pictures of the physical reentry, but what about the psychological reentry? It is not unique to astronauts. I guess you know it occurs probably to everybody who has a peak experience. Afterwards, what do you do, because you are on the ground for the rest of your life? How do you deal with this psychologically? Maybe it goes beyond just the experience of the person who has been there, and that is why I want to share it at this conference. This room is filled with people who dream of space, people to whom space is important. Even though most people in this room will never go into space, our ideas of what will happen in space in the future are very important to us. Yet we all recognize, in moments of calm reflection, that probably we will never see most of these things happen. We have to spend our real working lives in a much more mundane situation. How do we actually deal with it? The day before my flight, I met with members of my family, and one of my brothers gave me a piece of paper and said "*Just before you are ready to come back to Earth, if you are feeling sad about having to give it all up, open this up and read it*". And I did. It wa#o that maybe a little bit more than before some of us, although we cannot "see", can at least "know". Thank you.

(*) Original in French.

Intervention

Eric M. Jones

Dr. Hoffman said that we may not be prepared psychologically to leave the sun and point sources of gravity. I notice that in settlement experiences of the past, many people have discovered that the reality is not acceptable and they have gone back home.

LE STATUT PHILOSOPHIQUE DE LA TERRE EN QUESTION

"Mais ceci, avoir été une fois -même si ce ne fut qu'une fois, avoir été de cette terre, cela semble irrévocable".
R.M. Rilke, 9ème Elégie de Duino

Maria Villela-Petit
chargée de recherche en philosophie
C.N.R.S. Paris
59 rue Lhomond, 75005 Paris, France

Dans son ouvrage La crise des sciences européennes et la phénoménologie transcendantale, Edmund Husserl fait référence au génie à la fois décrouvrant et recouvrant de Galilée (entdeckender und verdeckender Genius) [1]. Sous couvert du nom propre de Galilée ce qui est visé c'est la science moderne elle-même, souvent justement qualifiée de galiléenne, pour autant que c'est à l'oeuvre du grand florentin qu'elle doit l'impulsion décisive de sa constitution. Plus essentiellement encore, ce n'est peut-être pas seulement la science moderne, mais la structure même de la vérité qui, à chaque fois dans sa forme épochale, comporterait une dimension de dévoilement et une autre de voilement.

Quoi qu'il en soit de cette dernière et immense question concernant la structure de la vérité, c'est en tout cas en faisant fond sur une compréhension de la science moderne comme à la fois découvrante et recouvrante, que je me propose d'envisager l'avènement, dans la pensée des Temps Modernes, de quelque chose de remarquable, quoique peu remarqué : le recouvrement, l'occultation de la singularité de la Terre.

Cette affirmation peut paraître, à première vue, tout à fait contestable. Car n'est-ce pas dans les Temps Modernes et dès l'aube de ces Temps, grâce au formidable essor des voyages maritimes rendant possibles de nouvelles découvertes et conquêtes, que se fait peu à peu l'appropriation empirique, par l'homme européen, du globe terrestre dans son ensemble ? Cela n'est que trop vrai. Et pourtant il n'y a pas de paradoxe à ce que, d'une part, l'homme européen se soit, pour ainsi dire, approprié l'étendue terrestre, en ait fait le tour, en permettant par là, entre autres, la constitution d'une nouvelle géographie, et que, d'autre part, la singularité de la Terre se soit voilée, voire occultée. S'ils peuvent être subrepticement liés, ces deux phénomènes ne sont pas du même ordre. Nous avons d'un côté un phénomène d'appropriation et de connaissance d'ordre empirique, encore qu'il soit gros de retentissements théoriques, et de l'autre un événe-

J. Schneider and M. Léger-Orine (eds.), Frontiers and Space Conquest, 209–219.
© 1988 by Kluwer Academic Publishers.

ment dont la portée est d'ordre à la fois scientifique et métaphysique, quoique les conséquences en aient pu être, comme nous le verrons, tout à fait concrètes. Disons que l'on aurait pu avoir Colomb sans avoir nécessairement Galilée et Descartes, même si ce sont leurs accomplissements pris ensemble et la constellation qu'ils forment qui déterminent cet âge que l'on appelle Moderne ou ces Temps qu'on appelle Nouveaux, temps qui pour nous relèvent déjà du passé [2].

Pour rendre plus explicite ce que nous venons de suggérer, interrogeons ce qui est ici en cause, à savoir ce que nous sommes en train de désigner par occultation de la singularité de la Terre. Deux questions s'imposent :

- à quoi cette occultation de la singularité de la Terre est-elle due ?
- qu'est-ce qui rend possible, aujourd'hui, d'entrevoir et de commencer à rendre visible cette occultation en tant que telle ?

C'est à travers la tentative d'apporter une réponse à ces questions que s'éclaircira du même coup pourquoi il importe de penser la Terre dans sa singularité.

Il est à noter que si j'évite de "faire image" en utilisant la métaphore de l'éclipse, c'est qu'alors que dans un phénomène comme celui de l'éclipse l'obscurcissement est spectaculaire et, par conséquent, vu comme tel, l'occultation de la Terre à laquelle je me réfère ne se remarque même pas, ou presque pas. On a du mal à la voir, et surtout du côté de la science.

Mais venons-en à notre première question.

1. A quoi l'occultation de la Terre a-t-elle été imputable ?

Moins, comme on serait tenté de le croire, au décentrement copernicien, c'est-à-dire au décentrement astronomique de la Terre par rapport au Soleil, qu'à l'émergence, avec la physique galiléenne et newtonienne, d'une nouvelle conception de la nature. Cette nature, l'homme cherche à la connaître dans la généralité de ses lois mécaniques, lesquelles s'imposent désormais à tout ce qui est de l'ordre de la nature sans exception. La nouvelle science présuppose ainsi l'unification et l'uniformisation de la nature. Car seules cette unification et cette uniformisation l'approprient à l'homogénéité de l'espace euclidien, en la rendant par là conforme au projet de "reductio scientiae ad geometriam". A la lumière donc de la science moderne, la Terre non seulement n'est plus qu'une planète parmi d'autres (elle est "mise à sa place", a-t-on pu écrire), mais encore elle ne se laisse appréhender par le savoir que fragmentée, morcelée en une multiplicité de phénomènes, lesquels pris en eux-mêmes et dans les relations qu'ils entretiennent les uns avec les autres doivent être subsumés sous des lois générales. Or la singularité de la Terre n'est pas de l'ordre de l'objectivable et n'est donc pas susceptible d'être subsumée sous des lois. Par là même elle échappe à une considération d'ordre scientifique.

(Une telle situation, remarquons-le, se trouve aujourd'hui bouleversée du fait même des développements que connaissent les sciences physiques et, en particulier, l'astrophysique, lesquelles nous amènent à comprendre, entre autres, que l'Univers a une histoire. Or de cette histoire une des étapes décisives a dû être celle où ont eu lieu les réactions physiques rendant possibles les transformations qui ont conduit à l'apparition de la vie. Cette nouvelle vue de l'univers a entraîné et entraîne, bien entendu, le surgissement de nouveaux paradigmes cosmologiques, susceptibles d'apporter une sorte de réhabilitation cosmologique de la Terre. Mais n'allons pas trop vite. Pour le moment, il convient de nous tenir à une considération de la Terre, de ce que j'appelle son occultation, à la lumière de la compréhension de la nature qui était celle des Temps Modernes.)

En résumé nous dirions que c'est la généralité des principes et des lois physiques, à travers lesquelles l'unité de la nature était postulée, "qui voila", alors, la Terre dans sa singularité. Du même coup cessa-t-elle d'avoir le statut d'un "singulare tantum" qui avait pu être le sien dans la cosmologie ancienne et médiévale.

Mais qu'a pu signifier la perte d'un tel statut qui somme toute ne semblait correspondre qu'à une cosmologie et à une science périmée ? Disons en une première approche qu'elle a signifié ceci : on cessa de voir qu'en tant que milieu de vie, en tant qu'<u>en</u> elle et <u>avec</u> elle s'est constituée cette trame de vie que nous appelons justement <u>terrestre</u>, la Terre est unique.

Que l'on me comprenne bien. Avec un tel propos je n'exclus nullement la possibilité qu'il y ait d'autres astres où la vie ait pu surgir et se développer. L'universalité des lois naturelles me l'interdirait. Il se pourrait, en effet, que quelque part la Terre ait une planète soeur, ou plutôt une cousine lointaine que nous les terriens avons de plus en plus hâte de détecter, en vue, le cas échéant, d'entrer en contact avec ses habitants [3]. Mais une telle "cousine", si elle existait, ne serait probablement pas identique à la Terre, n'aurait sans doute pas la même carte génétique qu'elle, ni le même âge. L'hypothèse, qui relève encore plutôt de la spéculation [4] que d'une attente raisonnable en voie de confirmation, laisse donc intacte la question de la singularité de notre Terre. Si j'en crois, par exemple, "The Anthropic Cosmological Principle" de John D. Barrow et Frank Tippler, la Terre aura été le seul endroit de notre Galaxie à avoir permis l'apparition de la vie intelligente [5]. Enfin, il se pourrait que, grâce à une colonisation terrienne, une autre planète devienne habitable, porte en elle un fragment de l'humanité, mais dans ce cas nous serions devant un phénomène d'un tout autre ordre que le précédent, puisqu'il ne s'agirait alors que d'une extension de l'habitat terrestre.

Suspendons cette digression et revenons aux Temps Modernes pour les regarder cette fois du côté de la philosophie. Comment la philosophie assuma-t-elle ces changements épistémologiques qui n'étaient pas internes à la science, mais qui, dès le départ et dans leurs racines les plus profondes, se jouaient dans l'entre-deux, dans l'inter-face, de la science et d'elle même ?

Pour faire bref, disons que face à la nature, lieu du déterminisme des lois causales mécaniques, lesquelles avaient rendu caduque la polarité ouvrante du ciel et de la Terre, la philosophie alla poser une non-nature.

Qu'en est-il de cette non-nature ? Elle s'oppose principiellement à la nature, voire en est séparée par un gouffre, dans le mouvement même par lequel elle s'objecte la nature, la pose en vue frontale, comme dirait Maurice Merleau-Ponty. Selon les grands systèmes philosophiques post-newtoniens, et leurs accents respectifs, la non-nature est pensée tantôt comme liberté, Sujet ou Moi transcendantal, tantôt comme Esprit. Termes qui, quoique distincts sont, alors, en partie interchangeables, et qui tous, d'une façon ou d'une autre métamorphosent la substance pensante de Descartes, la "res cogitans" dans son opposition radicale à la "res extensa", à la substance étendue.

C'est donc entre non-nature et nature que se distribuent les grands couples conceptuels qui organisent le pensable dans les Temps Modernes. Du point de vue épistémologique est nature ce qui se laisse objectiver et connaître par le Sujet, dont le projet de connaissance opératoire est, par nécessité interne, inséparable d'un projet de maîtrise. Or, dans l'horizon d'un tel affrontement entre non-nature et nature -celle-ci dans le cas les plus extrêmes, comme celui de la dramaturgie fichtéenne, n'étant plus vue que comme l'adversaire à vaincre, à rendre esclave [6]-, ce n'est pas seulement le statut de la Terre qui s'obscurcit, c'est aussi et peut-être avant tout celui du corps propre, du corps vivant (V. en allemand la distinction entre "Leib" et "Körper") qui n'a plus lieu d'être, qui n'a plus de lieu où être. Le corps aussi est "mis à sa place", refoulé du côté de la nature à objectiver par un Sujet purifié, au moins transcendantalement, de ses adhérences cosmiques, lesquelles sont d'abord ses adhérences terrestres.

On aurait deviné qu'une telle constellation de pensée ne pouvait que susciter de puissants malaises, y compris ce que l'on a pu appeler le "malaise" romantique, lequel fait alors figure d'une protestation aussi grandiose qu'inefficace. Parmi les grands penseurs idéalistes, Hegel, en particulier, va tenter, dans la construction de son système, de dépasser dialectiquement l'opposition entre nature et non-nature dans ce qu'elle a de plus heurtée, de plus dualiste. Il n'empêche : c'est toujours la même opposition qui règne et elle n'autorise pas une véritable appropriation pensante ni de la question du corps ni de celle de la Terre.

Et pourtant cette appropriation pensante va faire son chemin en philosophie ; d'abord latéralement dans l'oeuvre d'un penseur comme Nietzsche, ensuite dans l'oeuvre phénoménologique de Husserl et dans celle de Maurice Merleau-Ponty, ou encore, quoique différemment mais peut-être aussi plus décisivement, dans celle de Heidegger. Toutefois, en dehors même de l'espace philosophique, il nous importe de reconnaître comment la question de la Terre est devenue un des

enjeux les plus décisifs de et pour notre civilisation.
Nous entrons donc dans le champ de notre seconde question, telle que nous la formulions ci-dessus :

2. Qu'est-ce qui rend possible aujourd'hui d'entrevoir l'occultation de la singularité de la Terre, comme telle, c'est-à-dire comme une occultation ?

Permettez-moi de placer en exergue à cette seconde Partie de mon exposé quelques paroles prononcées en 1854 par un chef indien Seatle : "... Toutes choses se tiennent, disait-il,... Nous savons au moins ceci : la Terre n'appartient pas à l'homme, l'homme appartient à la Terre. Ce n'est pas l'homme qui a tissé la trame de la vie ; il en est seulement un fil. Tout ce qu'il fait à la trame, il le fait à lui-même."

De telles paroles, on s'en serait douté, ne parlaient pas, ou si peu, aux contemporains de notre chef indien. Que s'est-il entre temps produit pour que de nos jours elles puissent nous interpeller, et qu'on les entende à la fois comme primitives et en avance sur notre temps ?

En guise de réponse contentons-nous d'un bref rappel, ou plutôt d'une simple énumération : accidents chimiques, nucléaires, disparition accélérée des espèces végétales et animales, dégradation de la couche d'ozone... Et puis, comme si elle planait au-dessus de tout cela, la menace de l'hécatombe nucléaire. Voilà ce qui nous rend aujourd'hui sensibles à la protestation de cet indien du XIXème siècle, reprise depuis lors par d'autres voix. Voix qui ne parlent plus avec un accent primitif, mais qui sonnent tout simplement clairvoyantes et responsables. (Clairvoyance, je m'empresse de le dire qui n'a lieu que là où l'on ne s'enferme pas dans la nostalgie phantasmatique d'un quelconque paradis pré-scientifique).

Remarquons comment notre situation diffère de celle décrite par Pascal dans le célèbre passage où l'homme s'éprouvant, conformément au paradigme métaphysique des Temps Modernes, dans son étrangement au sein d'une nature devenue inhospitalière et muette, se voit, se pense comme un roseau pensant. Certes un rien de la nature peut toujours suffire à anéantir l'homme. Malgré les extraordinaires progrès de la médecine, la vie est toujours un équilibre fragile et précaire. Ce qui toutefois est nouveau, c'est qu'une erreur humaine peut aussi, de nos jours, suffire à détruire, non certes la "nature", mais cette Terre, qui est pour nous la nature, celle dont l'humanité toute entière ainsi que l'ensemble du monde vivant dépendent pour vivre. Autrement dit, ce qui est nouveau c'est que la Terre ne soit plus "indestructible et infatigable", comme la chantait le Choeur de l'Antigone de Sophocle. Cette Terre (Gè , Gaïa) que les Grecs, Platon encore y fait écho, tenaient pour la plus ancienne et vénérable des déesses sous le ciel... Et pourtant depuis Homère les Grecs savaient déjà, à la différence des Amérindiens auxquels je faisais tout à l'heure allusion, que l'homme est aussi celui qui "tourmente" la Terre, inscrit en elle les produits de sa technè, de son art [7].

Chez eux, néanmoins, un tel rapport tensionnel de l'homme avec la Terre s'accordait avec la vénération, la reconnaissance respectueuse. En témoigne la façon même dont étaient édifiés les temples consacrés aux Olympiens, qui avaient supplanté le sacral de la Terre-mère. Ces temples étaient situés de telle sorte qu'ils regardaient respectueusement vers les montagnes, symboliques de la sacralité de la Terre [8]. En outre, si Gaïa était un singulier, et "phusis" un nom commun, certains des traits signifiants qui parlaient encore dans l'entente grecque du mot "phusis" renvoyaient la compréhension de celle-ci à une expérience de la Terre dans sa force germinative. C'est en ce sens qu'Aristote pouvait penser la "phusis" dans son opposition à la "technè", à la production instrumentale ou artistique [9].

Mais que peut encore être la Terre pour nous aujourd'hui ? Nous ne saurions certes plus la vénérer comme une déesse. Pourtant nous ne pouvons assurer une chance de vie aux générations virtuelles qui sont appelées à nous succéder que si nous savons la ménager, la sauvegarder. Que si, par un tournant radical dans nos modes de pensée et d'agir, nous savons nous reconnaître dans notre réelle dépendance de la Terre, de la Terre dans sa singularité de Terre. Celle-ci n'est pas une simple, et somme toute infime, partie de la nature, de l'univers, partie dont nous pourrions nous passer après l'avoir exploitée jusqu'à l'épuisement. Dans sa singularité, elle n'est pas une juxtaposition de parties isolables et autonomes et déjà parce qu'en elle, telle que nous la connaissons, l'organique et l'inorganique ne sont pas dissociables [10]. Ou encore : tout en étant une infime partie de l'Univers, elle a l'insigne privilège de porter la vie et la vie de ceux qui peuvent connaître l'Univers. En bref, et pour le dire d'un mot, l'humanité et la Terre ne sont pas isolables, séparables, elles s'entre'appartiennent.

A vrai dire, la désoccultation de la singularité de la Terre commence déjà un peu partout à être ressentie comme une urgence. En ce sens saluons des ouvrages tels que celui de Jonathan Schell, intitulé "The Fate of the Earth", qui, tout en traitant la question du surarmement et de la dissuasion nucléaire, fait preuve d'un authentique sens philosophique en ce qui concerne le dégagement du statut de la Terre. Du côté de la philosophie, j'ai déjà évoqué, en passant, les noms de Husserl et de Heidegger qui ont posé quelques jalons essentiels pour une pensée de la Terre. Husserl, parce que l'on trouve dans son oeuvre, des éléments pour une élaboration plus fine de la question du corps propre dans son articulation avec celle de la Terre, en tant que sol (Boden, Grund). Heidegger, parce qu'il nous met sur le chemin d'une pensée de l'habitation, en nous faisant en même temps entrevoir comment les traditions métaphysiques qui ont dominé l'Histoire de l'Occident ont tendu à l'oblitérer.

Il nous reste à présent, et pour finir cette communication, à considérer le statut de la Terre en fonction du thème de notre Colloque : "Frontières et Conquête spatiale". A première vue, la réalité de la conquête de l'espace extraterrestre, les rêves et les motivations qui l'animent ainsi comme l'imaginaire qu'elle suscite sont susceptibles d'apparaître comme un obstacle supplémentaire et de taille à cette réappropriation du sens de la Terre que nous postulons comme une des tâches intellectuelles et spirituelles les plus urgentes pour notre civilisation. Bon nombre d'expressions et d'arguments utilisés par les partisans enthousiastes de cette conquête pourrait en effet nous le faire craindre. Nous n'assisterions alors qu'à un renforcement de l'occultation de la singularité de la Terre, comme conséquence inéluctable de l'avancée technique permettant à quelques-uns d'entre nous d'entreprendre des voyages extraterrestres. Il ne fait pas de doute que le danger existe de voir la conquête spatiale servir d'alibi à la poursuite de la dérérioration du milieu terrestre. Sans même parler de la question militaire (et pourtant il faudrait aussi en parler...), la tentation est grande d'envisager cette conquête comme offrant des possibles solutions de rechange pour une Terre devenue bonne à jeter, à rejeter, preuve de la permanence dans nos rapports avec elle du schéma sujet-objet, schéma métaphysique, que la phénoménologie essaie d'ébranler en montrant son pouvoir d'occultation, d'oblitération. Tout se passe d'ailleurs, au moins au niveau de certains discours de vulgarisation, comme s'il y avait des lendemains qui chantent sur quelque paradis extraterrestre... Et une telle idéologie est bien là, puisqu'elle se laisse aisément lire dans les scénarios qui nous décrivent les étapes possibles de la nouvelle conquête, voire des "colonisations" en perspective...

Toutefois les choses sont bien plus complexes. Et le réel nous surprend toujours... Les voyages extraterrestres, les séjours en station orbitale non seulement seront l'occasion d'une exploration élargie de l'univers, une exploration non entravée par l'obstacle atmosphérique, ainsi que d'un élargissement de notre champ expérimental pour l'obtention de nouveaux matériaux, de nouveaux médicaments, etc..., mais, peut-être, nous réserveront-ils aussi la surprise de nous faire mieux comprendre jusqu'à quel point le corps humain est un corps terrestre. Par là même, la sortie de la Terre aura été la façon détournée de mieux "la" découvrir dans ce qui la constitue en propre, de lever l'occultation qui nous empêche de la voir dans sa singularité, dans sa beauté. Nous serons alors mieux à même de reconnaître ces mille liens qui nous unissent à elle, et qui inscrivent notre appartenance à la Terre, dans sa conjonction particulière avec le Soleil, au plus profond de notre être. Et, encore une fois, ces liens ne concernent pas seulement nos corps (si nous entendons ceux-ci simplement comme des organismes), mais aussi nos modes divers de parler, nos langages (avec leurs métaphores et analogies) et, par conséquent, nos manières mêmes de penser.

Enfin, et pour conclure, osons espérer que la sortie hors des frontières de l'espace terrestre, que cette "conquête spatiale" dont nous ne savons pas encore les limites, soit le détour indispensable de notre réconciliation avec la Terre, réconciliation sans laquelle aucune réconciliation de l'humanité avec elle-même ne semble envisageable.

Notes :

[1] Cf. Edmund Husserl, Die Krisis der europäischen Wissenschaften und di transzendantalen Phänomenolgie, Husserliana B. VI, M. Nijhoff, 1962, trad. fr. par G. Granel, La crise des sciences européennes et la phénoménologie transcendantale, Gallimard, 1976, en part. le § 9, "La mathématisation galiléenne de la nature".

[2] Grosso modo, pour ce qui est de la science, nous pourrions baliser ces Temps Modernes par, d'un côté, Gallilée et, de l'autre, l'achèvement de la physique newtonienne (v. l'oeuvre d'un Laplace, par exemple) ; avec Faraday et Maxwell amorçant les changements qui rendront possible l'avènement d'un autre âge de la Physique. Pour ce qui est de la philosophie, les Temps Modernes se déploieraient entre Descartes et le post-hegelianisme. On ne saurait oublier que des continuités et des discontinuités marquent aussi bien l'histoire de la science que celle de la philosophie.

[3] Sur l'état de cette question et les moyens techniques dont on dispose en vue de "la recherche de la vie dans l'univers", on consultera The Search for Extraterrestrial Life: Recent Developments, Symposium de l'Union Astronomique Internationale, n° 112, Papagianis, Editeur : Reidel, 1985 et La vie dans l'Univers, J. Heidman, Fayard (à paraitre).

[4] En fait, la recherche de la "vie extraterrestre" a derrière elle tout un passé de spéculations philosophiques qui mériterait d'être mieux connu et interrogé, y compris dans ses racines théologiques. En se fondant sur la toute-puissance du créateur, John Locke, par exemple, déduisait la probabilité de vies "intelligentes" extra-terrestres bien supérieures à celle de l'homme. Voir sur l'histoire de cette spéculation dans les Temps Modernes le récent ouvrage de Michael Crowe, The Extraterrestrial Life Debate 1750-1900. The Idea of a Plurality of Worlds from Kant to Lowell, Cambridge University Press, 1986.

[5] De telles questions restent bien entendu tout à fait ouvertes, et ne sauraient trouver des réponses qu'à travers la recherche empirique (pour laquelle l'installation d'instruments en dehors de l'atmosphère ouvre d'extraordinaires perspectives) et les calculs la préparant. Je constate seulement que la thèse de l'ouvrage de Barrow et Tipper concernant la non-existence de vie intelligente dans notre galaxie, ailleurs que sur Terre, semble pouvoir être raisonnablement avancée au sein de la communauté scientifique, même si elle y rencontre d'énormes résistances. Mais peut-on être au clair sur la nature de celles-ci quand on sait la part de rêve, de mythe et même d'idéologie qui ne cesse de féconder (et parfois d'égarer) l'imaginaire sientifique ?

Peut-être revient-il ici au philosophe de rappeler modestement que le "scepticisme" aussi est nécessaire à la pensée scientifique...

[6] Voir J.G. Fichte, Die Bestimmung des Menschen (1800), trad. fr. par M. Molitor, La Destination de l'Homme, Paris, 1965. Remarquons que, d'un point de vue humaniste, la pensée de Fichte peut être interprétée comme transférant à la nature la qualité d'esclave, qualité issue du champ des rapports sociaux. C'est désormais à la nature, et non à certains groupes humains, de devenir "esclave". Mais cette position ne comporte-t-elle pas une autre forme de méconnaissance ?

[7] Cf. Sophocle, Antigone, Str. 1, trad. fr. par P. Mazon, "Les Belles Lettres", 1981. *"Il est bien des merveilles en ce monde, il n'en est pas plus grande que l'homme. Il est l'être qui sait traverser la mer grise, à l'heure où soufflent le vent du Sud et ses orages, et qui va son chemin au milieu des abîmes qui lui ouvrent les flots soulevés. Il est l'être qui tourmente ("apotruetai") la déesse auguste entre toutes, la Terre, la Terre éternelle (indestructible) et infatigable, avec ses charrues qui vont chaque année sillonnant sans répit, celui qui la fait labourer par le produit de ses cavales (c'est-à-dire ses mules). ... Mais ainsi maître d'un savoir dont les ingénieuses ressources dépassent toute espérance, il peut prendre ensuite la route du mal tout comme du bien..."*

[8] Voir sur cette question le remarquable ouvrage de Vincent Scully, The Earth, the Temple and the Gods, Yale University Press, rev. ed., 1979.

[9] Ayant cette opposition en vue, nous pouvons comprendre pourquoi il ne serait pas venu à l'esprit d'un Aristote d'employer la métaphore ou le paradigme de la "machine" pour se référer à la Terre ou à la nature (phusis). Or cette représentation est devenue pour nous tout à fait habituelle. Nous ne trouvons rien d'étonnant à ce que, par exemple, dans le "Grand Atlas Universalis de l'Astronomie", le chapitre sur la Terre s'intitule : "La machine Terre". Déjà dans le chapitre V du "Discours de la Méthode", Descartes l'avait employé à l'égard des animaux -les animaux-machines-, et nous parlons souvent de cette "merveilleuse machine" qu'est le corps humain. On ne saurait nier la portée heuristique d'un tel paradigme. Toutefois, ne devons-nous aussi remarquer la tendance que nous avons à biffer l'écart qui sépare le paradigme ou le modèle de la chose elle-même ? Et cette tendance ne risque-t-elle pas d'être égarante, lorsqu'elle empêche de penser la différence entre le modèle dont on se sert pour connaître et la chose qui est à connaître ?

[10] Voir, entre autres, le cycle terrestre de la production de l'oxygène, qui passe par la photosynthèse. Autrement dit, la Terre n'est pas seulement de la matière inorganique, mais le lieu d'échanges incessants entre l'organique et l'inorganique.

Interventions

Jean Heidmann

Je voudrais amplifier la question, qui a déjà été abordée le premier jour. C'est pour dire que "l'intelligence", à laquelle on se réfère si souvent ici, ne pourrait qu'être une étape intermédiaire. L'intelligence, dans son état actuel, n'existe que depuis quelques millions d'années (en étant large), et ce, en se limitant à la Terre ; qui elle n'existe que depuis 5 milliards d'années, le dernier de la phase actuelle d'évolution de l'univers, commencée il y a 15 milliards d'années par le Big Bang. C'est dire que l'intelligence est très récente, et très locale. Or, pour les cosmologistes, l'univers peut poursuivre sa carrière, pratiquement dans le même état, pour une centaine de milliards d'années. D'autre part il existe au moins cent milliard de galaxies avec chacune des milliards d'étoiles. Il est alors impensable de penser (oui !) que "'intelligence" soit le nec plus ultra de ce que l'évolution du cosmos puisse engendrer. Que peut-il exister au-delà de l'horizon de "l'intelligence" ? On en est à de la pure spéculation. Cependant les astrophysiciens-cosmologistes voudraient faire passer aux philosophes le message selon lequel "l'intelligence" est un concept très restreint dans son application à l'univers dans lequel nous vivons. Ensuite, et en ouverture à des progrès futurs éventuels, il faut signaler que si les efforts actuellement déployés pour aborder observationnellement la réception de signaux extraterrestres "intelligents" éventuels, en cas de réussite nous obtiendrons de 'linformation sur cet au-delà de l'horizon de "l'intelligence" dans l'univers.

Bernard Foing

- A propos du "principe cosmologique anthropique", il faut préciser que si les calculs peuvent paraître sérieux, la faiblesse vient des hypothèses de base qui peuvent être critiquées. Il faut rester vigilant devant les extrapolations et les "autorités", même scientifiques. Le problème de l'existence d'êtres "intelligents" dans l'univers devrait donc être considéré comme ouvert.

- Si l'évocation de Descartes et de Kant comme exemples de réflexion sur la science de leur temps m'apparaît tout à fait pertinente, je voudrais que soit posée la question de la position de la philosophie d'aujourd'hui par rapport aux sciences du XXe siècle qui ont fait jaillir de nouveaux concepts et problématiques : physique, science de la vie, étude de systèmes complexes, sciences de l'homme. Il y a une responsabilité philosophique à diffuser leurs questions vers la communauté.

ON THE INTERRELATION BETWEEN TECHNOLOGY AND EVOLUTION

Rudolf Albrecht (*)
Space Telescope European Coordinating Facility
European Southern Observatory

*) Affiliated to the Astrophysics Division,
Space Science Department, European Space Agency.

Abstract

Assuming evolutionary principles as providing the motivation for human actions, we demonstrate that the development of technology in order to extend habitable space has had positive evolutionary value. Extrapolating the current growth of technology and associated knowledge, and judging from developments in information science, we conclude that technology and knowledge will exceed human capacity to handle them efficiently, and will at the same time become disassociated from humans, rendering current evolutionary goals obsolete. Possible long-term scenarios are examined.

Introduction

The history of mankind is dominated by the quest for resources in order to improve the chances of survival of the species. This quest for resources also led to the development of technology, with spaceflight being the most thorough application. Thus, the present global motivations, goals, and means to achieve them can be explained entirely in terms of the theory of evolution. Other authors [1] have shown that this extends to mankind's system of values and intellectual processes.

Recently concern has been raised about possible negative impacts of scientific-technological developments on the medium-to-long evolution of the human race. Modern medicine, for example, is increasingly successful in keeping individuals alive who would have been eliminated from the evolutionary process only a few generations ago [2]. Effects range from a proliferation of individuals with impaired eyesight (harmless) to a proliferation of individuals with impaired resistance to infectious deseases (not harmless).

It is evident that a global undertaking like the conquest of space, especially including its ramifications and spin-offs, will change the course of human evolution, probably drastically.

J. Schneider and M. Léger-Orine (eds.), Frontiers and Space Conquest, 221–227.
© 1988 by Kluwer Academic Publishers.

Evolution

It would exceed the scope of this paper to treat the subject of evolution in a comprehensive manner. We are also aware that simplistic concepts of the evolutionary mechanism are not sufficient to completely explain all possible aspects. However, it is useful to summarize those elements which are considered valid for the purpose of this discussion and which will be important for the chain of conclusions.

Evolution (as opposed to development) is not goal-oriented, it has no "direction", except that big jumps in complexity are not possible. Thus, more complex systems have to be preceeded by less complex systems, simulating an "upward" gradient, with the appearent "goal" of achieving perfection.

It is important to note that abilities evolved to cope with certain boundary conditions in general will change those conditions. The ecological niche (or, if we include intellectual abilities, the evolutionary niche) evolves along with its occupants. This means that there are no "right" or "wrong" ways to cope with boundary conditions: what was right in the past might be wrong now. It also means that an unsuccessful evolutionary path cannot be re-traced.

Evolution of a species can continue along an equable path as long as the boundary conditions are steady and as long as the behaviour of the most significant fraction of the individuals is teleonomous, i.e. directed towards passing on to their offspring those abilities which in turn enable them to more successfully reproduce. Clearly, stringent boundary conditions do not tolerate non-teleonomous behaviour. As the evolutionary niche of mankind gets more comfortable, we observe a shift from common goals (at the expense of the individual) to individual goals at possible common expense.

A major point is that there is no difference in the principles and mechanisms which govern biological and cultural evolution. In fact, our humanitarian system of moral values is a relatively recent product of evolution, and is only now really taking shape. Evolution knows no "good" or "evil", and when war in an appropriate teleonomous behaviour it will be employed [3].

Technology

It is commonly accepted that mankind acquired the dominant role on this planet throught the application of technology. Evolutionary, the ability to produce and use technology is the result of curiosity, which in turn was necessary to search for sustenance. Helpful in the evolution of technological abilities was the concurrent evolution of organizational skills, which are the foundations of our scientific/technological society.

Technology provides a way to acquire information about the environment without having to go through the tedious process of encoding it into the genetic material. The ramification is that this body of information can, and does, grow much faster than if the information had been gained through genetic evolution. This process started only comparatively recently (about 5000 to 10000 years ago), but already it has reached a stage when most of us could not survive for long without technological help [4].

As can be witnessed in everyday life, mankind is quite happy to utilize technology without proper (or even any) understanding. It is amazing to see even small children using technological systems like computers, which were not possible to build, or indeed contemplate, a few generations ago. This naive utilization of entities, the pragmatic attitude of taking something at face value, appearently allowed our ancestors to function in an environment which they could not understand.

Space

The motivation which causes mankind to conquer space is basically the same one which caused early man to explore and control their environment: the search for resources in order to improve the chances of survival of the race. The conquest of space needs the application of all technological means which have been developed.

There is one major difference between space projects and other technological projects: the lowest limit in size and complexity is considerable for space projects, and enormous for manned space projects. It is not possible to start small.

The history of the US Space Shuttle program with its wrong assumptions, schedule slippages, cost overruns, and final disaster teaches a dramatic lesson: the project has become too complex, it has outgrown the ability of humans to maintain a proper overview, assess the ramifications of changes, and anticipate operational needs. For some time it was possible to compensate for this by introducing management procedures which allowed large organizational entities to collaborate without a thorough understanding of each other's activities. This led to a vicious cycle, however: the management system itself, and, in particular, the associated bureaucratic procedures, became opaque, so they were not being followed correctly, or not applied properly [5].

This is not to say that we will not go into space. We will turn to technology for the rescue. We will use computers, in particular we will use the techniques known as artificial intelligence (AI) to remedy the problem.

Computers

Much has been written about the staggering growth of computer technology during the last three decades. This growth, however, has so far been mainly confined to improvements of the hardware. Most software improvements can be traced to improved hardware capabilities [6].

This situation is about to change dramatically. Already in the near future developments in computer science will render computer programming as we know it today obsolete [7]. Artificial Intelligence based Software technology will provide computers with natural language understanding, image recognition, and reasoning capability. Expert systems already allow one to de-individualize bodies of knowledge and make it available to the non-expert, or to another computer, on a 24-hour a day basis. Knowledge bases can be stored, merged, experimented with and checked for consistency. It will be possible to perform "what-if" operations, the usual hypothesis building tool used in science, in an automatic manner, on knowledge bases which are much larger than can be retained in the human brain. This will make true in the human brain. This will make true interdisciplinary research possible.

Along with these software developments we will see the advent of new hardware which will provide revolutionary man-machine interfaces, including direct brain-computer links. These concepts were first developed for the rehabilitation of accident victims through the use of electro-mechanic artificial limbs; impressive progress has been made. They are now being investigated for possible applications in military aircraft technology [8], [9].

Even very careful extrapolation shows that we are right now witnessing the beginning of a development which have enormous impacts on the further evolution of mankind. The original motivation to generate and use technology has caused mankind to create technological systems which will very soon outperform humans in many areas of intellectual activities: computers will be the better chess players, conversationalists, decision makers, and conquerers of space. To achieve all this, we do not even have to find out how the human brain works, it is just necessary to produce the same results.

Thus, computers, through Artificial Intelligence systems, will successfully help us to conquer space. Howewer, in the process of doing this, man will have become a passenger.

Impacts on evolutionary boundary conditions

Evolution's answer to drastic changes of the boundary conditions has been the elimination of the species concerned. For the further discussion we will assume that this will not be the case; in particular we assume that there will be no catastrophic changes of the boundary conditions through the ecological impacts of the application of technology, or by war.

Looking back at the history of mankind, the reproductive advantage has always been with those members of the species who had the most resources (stength, food, territory, money). Since resources are being gained with the aid of technology, evolution produced the technology oriented society which we are now.

There is a high probability that this will change as soon as technology becomes too complex for most people to understand, yet is easily accessible to everybody; when technological systems will be able to answer human questions, not necessarily correctly, but either to the satisfaction or beyond the understanding of humans.

As we do now, people will continue the naive use of technololy, even if they do not understand it. However, as humans are no more needed to produce and master technological systems, the sociological status of technology-oriented individuals and thus their reproductive advantages will decrease. Within a few generations the desire of mankind to conquer space may have vanished.

Extrapolations

Assuming the non-destructive development outlined in the previous section we may conclude that technology will ensure human physical survival as individuals by providing food, shelter, and other services. The question arises how the disasssociation of knowledge and technical skills from humans will impact the fate of the species.

One problem is that evolutionary paths in complex systems are quite unpredictable; another problem is that the "strategies" employed by evolution are not defined: the same goal can be achieved in a variety of different ways. Extrapolation is thus reduced to little more than speculation.

Given the fact that the decreased selection pressure in a comfortable evolutionary niche increases tolerance towards non-teleonomous behaviour, there will initially be a wider variety of human types, physically, psychically and culturally. Which one of those will usurp reproductive dominance is not predictable. By way of exclusion we can state that those who have lost the urge to engage in heterosexual contact will not. Neither will those that are being kept alive by extreme medical measures. Yet, not even this is certain, given the advances of genetic engineering and in-vitro fertilization [10].

For the sake of discussion let us assume that we evolve towards a society without the need to compete for resources, in which therefore the need for technical knowledge and skill vanishes because they are not required for the individual, if not discouraged because of the evident impossibility to compete with technological systems. At the same time it should be possible to experience all possible sensory peceptions by means of advanced man-machine interfaces. Will mankind use this for widening the intellectual horizon, or will this capability rather be used for eternal entertainement? And which group will pass their preference on to their offspring?

As already mentioned in previous sections, mankind will most likely continue to use technology naively, i.e. without feeling the need to understand how it works. It is interesting to note, however, that in the past there has been a tendency to populate the incomprehensible parts of the environment with supernatural beings (spirits, gods), which seems to provide a working hypothesis and a degree of control through worship and sacrifice.

There can be no definite conclusions from these considerations, except that straight forward extrapolation from the present situation is certainly not possible. The conquest of space will not be a re-play of the expeditions of the conquistadores, or of the exploration of the American west. From these exercises men emerged with their intellectual superiority intact and with the basic evolutionary boundary conditions unchanged. The conquest of space, through its implied technological advances, will challenge the very motivation of our civilization to continue the present course of development.

Acknowledgements

My wife Katalin Albrecht-Nagy contributed to this paper through many challenging discussions.

References :

[1] Popper, K.R., Lorenz, K., Die Zukunft ist offen, Serie Piper München, 1985
[2] Livingston, J.A., "One Cosmic Instant", Dell Publ., NY, 1974, p. 63
[3] Goodall, J., Bandenkrieg unter Schimpansen, Bild der Wissenschaft 8, 1985
[4] Lorenz, K., Der Verlust des Menschlichen, Serie Piper München, 1985
[5] Covault, C., Rogers Commission Report on Shuttle 51-L Loss Aviation Week and Space Technology 09.06.86, p. 16
[6] Albrecht, R., Next Generation Software Techniques, in SPIE Proceedings, "Instrumentation in Astronomy VI", Tucson, 1986, in press
[7] Bibel, W., Wissensbasierte Softwareentwicklung, in Informatik-Fachbericht 112 (Brauer, Radig, Eds.), Springer, 1985
[8] Simons, G.L., Expert Systems and Micros, NCC Publications, England, 1985
[9] Pfund, B., High-G Maneuvers, Aviation Week and Space Technology, 02.09.85, p. 98
[10] Walgate, R., In Vitro Fertilization, Nature, vol. 232, 10, 1986, p. 358

Interventions

Eric M. Jones

We became specialists disconnected from the production of our basic needs a long time ago. Part of the appeal of Daniel DEFOE's Robinson Crusoe is its description of the all of the skills and processes that support human existence but which DEFOE's urban readers know nothing about from personal experience. I do not believe that the increasing use of computers makes us superfluous, because I believe that most of what we do -whether we are urbun dwellers or nomadic hunters and gatherers- are social and symbolic activities. We will remain relevant because our relevance lies in our interactions with each other.

Jeff Hoffmann

- The strength of NASA's system is the ability of individuals can exercise creativity within a bureaucracy. If we lose this, the enterprise will not succed.

- I do not feel competition with computers. They help me. At the basis of my experience is the ability to do things human, to feel and to experience. That is what we value about the human experience and why we feel different on deep spiritual level when a human visits the moon from when a robot goes.

IS LIFE EARTH-LIKE?
Space flight, a most unusual experience

Wubbo J. Ockels
ESA Scientist Astronaut
ESTEC
Postbus 299, 2200 AG Noordwijk, The Netherlands

"T-36 seconds in counting, automatic sequence start". The Schuttle computers have taken over the countdown. I was sitting already over one and a half hour in the seat, lying on my back because the Shuttle stands on its tail. "So close to launch". Two years in Houston for the basic astronaut training. Three years training and preparation as a back-up for the first Spacelab flight. Now, 30th October 1985, sitting in the Challenger. I hope so badly that nothing will come in between me and the flight, that nothing goes wrong.

"T-6 seconds, main engine start", I feel the Shuttle getting alive. I count loud 3...2...1... and then "Yippie", the solids go, and there goes the Shuttle, there goes the D-1 Spacelab, there I go, to space; one feels the increasing speed. Three times the weight on Earth.

After two minutes, 40 km high, the two solids are burned out. With a big jerk they separate and then... nothing! As the Shuttle was already on an almost ballistic track, you feel no gravity anymore, only the acceleration of the Shuttle. This acceleration was only 1.2 g at that point in time, only slightly more than I used to feel the one-and-a half hours I waited on the pad, standing still! I thought for a fraction of a second that the engines quit. Fortunately the forces were building up again and a smooth six more minute ride brought us into Space. the engines stopped, we were weightless. the first I did was to take my pen out of my pocked and put it right in front of my nose. It stayed there "I could move my head arount it!" "We are weightless, we are in a no-force environment".

Very often I had tried to imagine weightlessness before. Now it was real and...different. One main question came back to me: "will I, being outside Earth get to think different about life?... Is life Earth-like? I think I found the greatest difference between space and Earth: the gravity, the inevitable gravity.

Life's evolution took place on Earth, man's concept of life was formed on Earth, how general is this concept?

The force of gravity to which everybody is exposed to all his life is always the same. I was striken by this new sensation of weightlessness, as if I had opened a door and saw a new colour I'd never seen before, but so real -I had only no name for it-.

J. Schneider and M. Léger-Orine (eds.), Frontiers and Space Conquest, 229–231.
© 1988 by Kluwer Academic Publishers.

Of course I had to stop this reflection and start to work. Over 70 experiments had to be done. Many of them as quickly after the launch as possible, because these experiments were to measure the onset of our adaptation to weightlessness. Some of us got sick, others enjoyed it. It was hard for me not to smile all the time. All of us worked hard and after two days we all felt great.

I had no time to think more about gravity and life till the work was done, 110 orbits around the world later. During seven days I got used to the feeling of freedom. No forces, no ties to a floor. When I relaxed and closed my eyes I could even forget my body. The purest feeling of rest. All the forces you felt resulted in movement, caused by an acceleration. I started to think again about man on the surface of Earth. Always pulled down by an incredible force, or better said, pushed upward by an enormous acceleration.

Each second everybody goes 36 km/h faster. That is 100 km/h in 3 seconds, 3 million km/h after one day, almost the speed of light after one year! The brain connot believe that of course. As the gravity force is always constant, we will ignore it. When we sit on a chair, we sit still. Horizontal forces, however, <u>are</u> interpreted as change of speed. When we feel a pressure on our back in a car, we go faster. So we developed a different look at horizontal versus the vertical. How far is a friend when looking down at him from a 100 meter high tower? And how close is he when you are on the ground 100 meter aways? How striking is the apparent increase in size of the Moon when he is at the horizon such that we apply our "horizontal" view, and how small does the Moon seem to be when looked at the zenit, straight above us? Is the brain fooling us? Of course! But how does this influence our concept of life, if it a all does? How would we think when perceiving gravity not as a static force, but as a real acceleration going faster and faster upwards? As I said I had been thinking about this often before the flight, but I could never really get myself in such a state that I felt gravity as an acceleration.

The de-orbit burn was completed and the slowed down Shuttle fell back into the earth atmosphere. Looking through the window saw the clouds come closer and closer. In credible became the speed at which they passed by.

Then, at about 200,000' the air drag started to build up. A deep orange glow developed around the nose. Quite sudden was the build-up of force, 0.8 g's. My head was pulled downward. It took a large effort to lift my arm. I felt at least four times as heavy as I remembered to be fefore the flight. It was uncomfortable. Strange effects occurred due to my screwed up vestibular system. Whatever I looked at moved when I moved my head. I did not like it! The ride down took 20 minutes.

The Shuttle touched smoothly the runway at Edwards Airforce base. A role-out and then stop, silence, but heavy, damned heavy. After we got the go-ahead from our commander, we unstrapped and exercised our leg muscles before trying to stand up out of our seats.

Now came one of the most impressive moments of my flight. I stood up and... it was frightening, for a few seconds I felt I just stepped into an elevator going up with and incredible speed. All my thoughts about perception of gravity came back immediately. I just perceived it as real acceleration! What does this gravity do to us? We are all so used to it that it must give a distorted view. We cannot escape it. The first few nights I waked up several times, switched to acceleration, the bed was pushing me upward, speed was building up. Then I switched back to the force feeling. I was tied down with hands and feet.

Somehow suddenly this inevitable gravity, its incredible speed and the going on of time continuously seemed to fit together naturally, time, seen as the necessity to experience changes, to live.

For all of us time goes with the same speed. To reach tomorrow, we all have to wait one day. As if we are on the same wave, forcing us towards the future... It makes me think of an analogy. On the other side of a lake as flat as a mirror sits a blind and deaf man. He fishes using his sensitive fingertips at the rod and noticing any movement of the water surface. If I want to communicate with him I can throw a stone in the water. The ripples will reach his float. The man will know that I am there.

Now suppose I am on a boat which moves at a speed equal to the speed of the ripples which disturb that flat water surface. I can throw as many stones in the water as I want, but the lake in front of the boat stays flat, undisturbed. I cannot communicate with it. If I want to see that part changed, I have to wait till I am there. Like I have to wait for tomorrow, yesterday is already disturbed, like the wake of the boat. If the analogy holds, you will ask yourself: "what causes the boat to go, where is the engine?".

Is the engine the gravity? Then gravity is giving us the speed at which information goes, like the ripples in the water. But that <u>is</u> the speed of light, 300.000 km/h.

What is then our speed <u>in time</u>? One hour per hour of course. This speed is <u>one</u>.

Is it then gravity, is it Earth which pushes us towards the future?

Is Earth giving us the sense of time?

Do we live the way we live because we are on Earth?

Is our concept of time and of life universal or does it depend on Earth?

Is life Earth-like?

I bet it is.

CONFINEMENT IN SPACE : THE HUMAN DIMENSION

Dr. Roberto Pinotti
Sociologist
Futuro srl, Florence, Italy

In 1979 the first U.S. space station SKYLAB descended in a fireball over Australia. Serious confinement problems in space manifested themselves at first during the short-lived SKYLAB program. In the opinion of NASA flight director Dr. Hutchinson, "the initial purpose of SKYLAB may have been to explore simply how to live in space, but the cost of the program -two and half billion dollars - caused us to change our minds... our system was designed to squeeze every minute out of an astronaut's day..." The efficiency of the first and the second SKYLAB crewmen made the flight controllers think this way, and so they decided to plan the third crew's day without leaving a spare minute, priding themselves that, from the time the men got up to the time they went to bed, every minute was programmed for them. The flight surgeons saw what was happening and tried to stop it, but it was useless. As Dr. Herry R. Hordinsky, the crew surgeon for the third mission, said later, "we witnessed Mission Control getting off on the wrong foot, but there was no place to blow the whistle". The main reason was not only communication difficulties between flight surgeons and flight planners during the third SKYLAB mission; but also the fact that nobody of the Mission Control was familiar with psychology problems. "At conferences, when we were on the side of easing up, of saying that the flight plans were too much, the engineers couldn't understand what we meant" Dr. Hordinsky said. Never before had so much data been assembled about a group of men for so much time. If SKYLAB was a test tube (this was the emblem of one of its three missions), the three crews were indeed tiny colonies isolated in it, to be observed and manipulated by the computers and men at Mission Control at Houston. "What nobody foresaw" The New Yorker staff writer Henry S.F. Cooper wrote in 1976 "was that some of the astronauts would rebel". And maybe this may have been one of the most important results of the whole SKYLAB experiments, during which the men had to report every mouthful of food they ate, and keep track of every urination and bowel movement.

The third crew went on strike at the end of the sixth week, when Gerald M. Carr, the commander, told the flight controllers -making a sort of declaration of independence to the ground- that he and his colleagues had enough. Of course at Houston some of the flight surgeons had noticed a progressive change in their characters. Lt. Colonel William R. Pogue, the pilot, and Edward G. Gibson, the science pilot, had grown thick, revolutionary-looking beards which had

J. Schneider and M. Léger-Orine (eds.), Frontiers and Space Conquest, 233–245.
© 1988 by Kluwer Academic Publishers.

made them uneasy. The third crew's remarks, at the same time, had begun to have a barracks-room grumpiness, expressing polemic and penetrating criticism. They didn't always refreshed in the morning, and sometimes they weren't eager to pop out of bed. the flight controllers had openly talked of these "reluctant subjects" as being lethargic and negative. More and more irritable, the astronauts complained that there weren't enough changes of clothes, that they didn't have a great enough variety of them and that they got tired of the dominating golden brown color in their clothing (caused by the fire-resistant material they were made of); besides, they had begun having insomnia: and to make matter worse, the Mission Control, instead of letting up on the schedule, actually began <u>increasing</u> the work load.

NASA had originally construed the word <u>live</u> in space to mean decently as one might on Earth, with regular shifts and time off for relaxation; but the Mission Control was under pressure from NASA administrators who wanted results and from the scientists whose experiments were aboard, and nobody realized the importance of the fact that the third mission was the longest in the SKYLAB program.

And so, one fine day, Carr, Gibson and Poque stopped working and did exactly what they felt like doing. "On the ground, I don't think we would be expected to work a sixteen-hour day for eighty-five days, and so I really don't see why we should even try to do it up here" commander Carr said. Actually, everybody except the third crew had lost sight of what was to have been the main purpose of the SKYLAB program, and that was just to see whether men could really live in space for long periods of time; and not to use men in space as the integrated components of an advanced technology. As a result, the entire SKYLAB third mission became a sort of affirmation of man over machine, and a lesson, one flight surgeon said later, that NASA was better learning late than never. Indeed, a few scientists and engineers at Houston regarded the astronauts' leisure time as an experiment in itself, the success of which would be essential to have men spend long periods of time in space stations, at moon bases or on three-year round-trip excursions to Mars. "At first, when they had begun asking for time off, I felt it was unwarranted" NASA flight director Hutchinson said later. "We now see that time off is mandatory. A man has to get mental enjoyment out of something other than his work. We now feel that an astronaut's time off must be inviolate. An I think that before the mission was over, we reached a reasonable compromise. But we learned a marvellous lesson in how to manage people". In fact, NASA had discovered that astronauts didn't like being relegated to the role of robots or of maintenance men, and they retorted that no remotely controlled device could really replace their abilities to take advantage of opportunities that came up suddenly or of new targets and situations to be faced from time to time. No machine can act creatively. "After the third SKYLAB mission, it is hard to see how astronauts can ever be the same again" The New Yorker staff writer Henry S.F. Cooper remarked. Since February 8th, 1974, when the third and final crew of American astronaut hopefully left SKYLAB closing down their space station in such a way that it could easily be put back in commission, NASA's interest in human behavior in space became a new challenge in the Eighties, developing another scientific approach in space research: the need to define the variety of human problems and reactions in space through social sciences. The Soviets have had space stations in orbit (Saluyt 1 through 7) since 1971, and placed a high priority on the mental well-being of their space crews by applying the support of psychologists to their long-term missions, in order to mitigate most of the problems caused by confinement.

In the United States, on the contrary, the early astronauts (and especially the "Right Stuff" Mercury team) had very little interest in psychological intervention, and there were two things they reportedly considered intolerable: rectal thermometers and psychologists. This witticism shows how in the U.S.A. the focus had been at first on technological engineering, owing to the relatively short duration of the missions. But the scene changed. Today the Soviets have 12 man-years of data on the effects of prolonged weightlessness on the human body and mind. Experience indicates that work performance may decline after stays of more than 4 months, due to gravity's heavy hand. Despite daily two-hour workouts on a treadmill and exercise cycle, Russian long-duration crews proved to be too weak to walk on their return and can find even the weight of a bed sheet quite uncomfortable. As with American astronauts, in flight about half the Soviet cosmonauts suffer motion sickness.

Both Russian and American long-duration space crews experienced hard work under rigorous and trying conditions. People who have lived in space small, crowded and austere stations report depression, restlessness and boredom. Russian cosmonaut Valentin Lebedev, for example, experienced growing fatigue, lack of sleep and thoughts of home after seven months in orbit. The psycho-sociological components of the micro-cultural universe inside space stations are often stressful. Soviet Salyut's daily routine includes arising for the crew at 08.00 hours, four meals with bread, assorted meats, cheese, honey, cake, prunes and coffee, two tea breaks, two hours of strenuous physical exercices against muscle atrophy and technical and scientific work until supper time. After the last meal in their daily activity, the cosmonauts visit with their families and friends on two-way television or talk with athletes and entertainers. These diversions are arranged by psychologists who oversee the mental health of the lonely spacemen in orbit. Then, at 23.00 they go to bed. That is, tucking into sleeping cocoons fastened to a wall, hoping for sound sleep. And this is really a rare luxury in an alien environment in which they can never entirely relax.

In the American SKYLAB station, for example, astronauts reported their disappointment for having been repeatedly awakened by others' night visit to the bathroom. With the absence of gravity it is easy for the cosmonauts to gain an inch or more in height. They have grown accustomed to the unpleasant atmosphere of the station but not to the awkward hygienic facilities, which substitute a vacuum for gravity in handling human wastes. As far as bathing is concerned, each physical exercise session generates on the cosmonauts' skin an envelope of sweat that it is not easy to towel off. Every ten days they shower, in a complex process lasting an entire day. Scientific work includes many biological experiments, especially in little greenhouses. Healthy vegetation could be essential for recycling air, water and wastes as well as for providing food for longer flights. Extra-vehicular activity is often a necessary task. Besides, constant high levels of machine noise along with routine sameness in work activities can be a serious source of stress. Just after the SKYLAB experience, NASA considered the astronauts' reaction to extreme isolation a very important psycho-sociological factor to face, too. In the next future, space habitats will have, after all, problems not too different from those found on Earth on isolated islands where like in outer space, though there is no lack of people in the immediate vicinity, thousands of miles of open void are beyond. And the awareness that you can't reach quickly and easily any other human being has deep psychological effects on the subjects involved, in which the dominating feeling is that of being alone, and not capable to cope with any serious emergency, since rescue will require a long wait. A wide spectrum of isolation and confinement effects manifested themselves in Antarctic stations, isolated islands and villages cut off for months by heavy snows. In this case a particular feeling of enclosure and entrapment, called the "Shiminagashi syndrome" or also, more commonly, "cabin fever", can ensue. As a consequence, a mild neurosis, a belief you will never be able to escape can result, eventually developing serious mental breakdowns of hysterical nature, most of which have often proved fatal in certain lethal external environments (because of the individual's uncontrolled attempts to leave the situation). On the contrary, we may have an opposite and often less noticed (and more dangerous) effect in a sort of steady, quiet withdrawal from human interaction. This has been observed frequently at Antarctic winter-over stations.

And this response -a withdrawal from reality as well- can cause unexpected risks through the misfunction of an important job, the breaking of safety regulations and other situations dangerous for their potential exposing all the inhabitants of these stations to the surrounding life-threatening environment.

Enclosure brings with it the possibility of crowding, a very important factor tending to aggravate the effects of other stressors. Reactions to crowded living range from poor health and poor learning to aggressiveness and alienation, and such highly detrimental effects were observed in different group situations. Edward T. Hall's studies on proxemics as a new approach in man's spatial relationships are very stimulating, and suggest the vital importance of the interaction of space with human subjects. Territoriality is important for animals as well as for human beings, and abnormal behavior in zoo animals may be caused not only by their forced confinement in cages, but also by crowding. As the result of Earth's evolution, man has similar problems. Findings from Antarctic research missions have shown that the need to avoid close contact can become quite pronounced before it is deemed unhealhy. Moments of privacy can be an effective way to recover psychological equilibrium. Besides, on Antartic research stations many people developed "Big Eye": they became mildly depressed and began having insomnia, while most of their metabolic functions became desynchronized. This was because people lacking normal daily cues of daylight and darkness tend, when allowed, to retire later and remain awake longer each successive night. And this practice, called "free cycling", caused a general lowering of both morale and productivity. In 1959, during the International Geophysical Year, six mentally disturbed people had to be removed from American Antarctic bases. On submarines, the changing of shifts in 24-hour operations proved to be so disruptive to crew members' sleep that the United States Navy had to develop a "sleep hygiene" program including regulating schedules and teaching techniques to ease going to sleep. On two cruises of one American Polaris submarine, approximately five percent of the crew of 137 experienced some form of psychiatric disorder, reportedly either depression or minor anxiety. A 1979 U.S. Navy research report found that of more than 22,000 men surveyed, 261 were psychologically disqualified while assigned to a submarine. And of these, 127 (48.6 percent) were from nuclear missile subs, which often submerge for a 90-day patrol. The lack of a private retreat for each crew member is frustrating. Tension, anxiety, depression, aggressiveness and other detrimental psychological effects were also reported in oceanographic research vessels, Artic construction and oil production camps, underground military command centers, supertankers, offshore oil rigs and experimental undersea habitats. All these situations, dominated by isolation and confinement problems, are surely comparable with long-term space flights. And since we know many of the negative effects they generated because of the absence of an a priori, definite psycho-sociological effort, in-depth comparative studies of the conditions man is going to face in space during future prolonged missions may prevent most of such detrimental effects to manifest themselves up there.

This was the chief cause of the establishment of a support team including social sciences experts for the Soviet cosmonauts. This "psychological support group" communicates with and monitors the Russian cosmonauts during the 15 to 20 minutes of each 90-minute orbit the Salyut is in contact with the Mission Control in Moscow. In addition to detecting and trying to prevent emotional disorder, the support group regularly sets up two-way televised meetings between cosmonauts and their relatives and friends. We must remember that the longest human space flight was 237 days on Russia's SALYUT 7, during which this psycho-social support proved to be of vital importance . The Soviet Union has logged 12 years of space travel, and the U.S.A. less than 5. Of course also NASA has grown more and more sensitive to design not only the physical, but also the psychological and organizational environment in orbit the American astronauts will have to live in. And the formation of their Space Human Factors Office at

Ames Research Center in California shows this more sophisticated attention of the United States space program to the psycho-social sphere, in the best tradition of the deep interest of the Americans for behavioral studies and social sciences in general. In the words of Dr. Yvonne A. Clearwater, environmental psychologist and leader of the Habitability Research Group of NASA's Space Human Factors Office, "our challenge is to troubleshoot the proposed space-station systems and settings while they are still being developed conceptually: first, identifying the environmental conditions likely to be significant stressors and disincentives; next, definying the kinds of psychological, emotional and behavioral problems these could produce; and finally, showing how such problems might affect work performance. We must demonstrate these potential costs before our recommendations for averting them will be accepted and integrated into the program design".

As a result, NASA is studying how people have responded to a range of comparable isolated and confined work settings, developing conceptual models, computer simultations and mock-ups to test, for example, relative trade-offs of variations in sizes and shapes of areas and rooms; interior layouts; furnishings; lighting; and color and texture of surfaces; and so on. All this in the light of the implementation of human productivity. According to NASA sociologist B.J. Bluth, " 'engineering logic' and 'elegance' does not include costs such as excessive fatigue, boredom, high work load, isolation and frustration on the part of the operators. The question is, 'does it work?' and not 'how does it look, or is it easy to use?'. There is a basic naivety in the assumptions underlying such a set of questions. A good human performance design is not simply a decoration. Good design 'works' for the operator, not just in the shop where it is made, in isolation from its context, but on the job. And it is possible to 'work' it day after day, hour after hour, effectively".

In other words, design and architecture are much more important for future space stations than it seemed to NASA administrators years ago. On the contrary, they can be the final answers to many psycho-social problems for man in space. This approach is reflected in Marc M. Cohen's station architecture studies at Ames Research Center, dealing with human factors issues as seen from an architect's eye.

As far the European Space Agency is concerned, another architect, ESA fellow Jérôme Gerber, developed a very interesting study on human factors in an analogous environment: the submarines of the French Navy. In his conclusions, he suggests the need of a Research Unit on Human Factors for ESA, and recommends European multidisciplinary studies to be carried on in this field.

At the same time, in Italy, another architect, Daniele Bedini, had formed in Florence a private, interdisciplinary study group of researchers aiming at developing a space habitat at its best. This was an effect of the general interest of Italian scientists for his original OLGA TOWN project for a 21st century geosynchronous space station, the potential of which had been acknowledged by NASA because of the unexpected practical benefits from the standpoint of both economy and habitability.

Due consideration was and is given by this Italian team to psycho-social implications, of course. As one of its members and a sociologist, I was a consultant for D. Bedini's recent paper presented at the 37th Congress of the International Astronautical Federation in Innsbruck dealing with "Space station habitability study: the relation between volumes, shapes and colours inside the space station and human behavior". It was our team's first approach to the space habitability new guidelines many researchers are looking for both in the United Stated and Europe. Our purpose is just to suggest possible human factor patterns for future missions in order to improve the cohesiveness of the astronaut groups through a correct social and organizational integration of the personnel aboard as a consequence of a general psychological equilibrium.

A part from the importance of Russian and American experiences and studies, we think too, as Europeans, that ESA ought to promote the setting up of a research unit on human factor as well, as Jérôme Gerber suggested last year, and we support this recommendation. His preliminary analysis on human factors as a consequence of his study resulting from his mission to the Toulon naval dockyard is surely a very good start in the light of ESA's studies for an European autonomous manned space station. All we may add are just some considerations of psycho-social nature.

First of all, neurosis in the subjects involved is a possible effect of isolation and confinement. And in neurotic subjects, according to the American psychiatrist Carl Rogers, the ideal and the real images of themselves do not coincide, and this is a definite cause of possible mental strain and discontent, affecting morale in general. Military discipline rules, without knowing all this, give a great importance to the formal look and care of soldiers, whose limited freedom and isolation can cause in them a similar form of neurosis. And in the confined situation of military life, a soldier's best key to the keeping of his real image is a very easy one: a mirror is enough. It is not a mere coincidence that stressed and neurotic subjects tend to manifest a general indifference for their personal look. And since the third SKYLAB crew had similar reactions, such possible detrimental effects can be limited by compelling the stressed astronauts to see their <u>real</u> images. In other words, a possible solution is what we may call the "mirror effect": that is, the use of specular surfaces inside the space habitat.

All this has not only a mental well-being effect, of course. As a social sciences researcher, I was surprised to discover that neither the Americans nor the Russians understood the importance of the use of mirrors inside a space station just to create an illusory but psychologically prominent enlargement of the space station's interiors: after all, a solution well known to architects in everyday life.

The symbolic value and the psychological role of different colors inside a space station was already faced in his IAF paper by my colleague D. Bedini. In any case, we agree with the idea of in-depht studies about human psychological reactions to different colours. The Soviets use differently coloured light suffused on walls as well as music inside "psychological relief rooms" experimented in industrial settings on Earth in order to increase productivity, and proposed similar relaxation techniques for their cosmonauts.

Another solution of the Russians consisted in covering the walls of the SALYUT 7's interiors with colored leather: a very good idea, since the use of natural materials reflects and satisfies the need of re-creating a warm environment, like cork does in submarines and leather and valuable wainscoting do inside luxury cars. If associated to "positive" colours, this can produc both individual and group stability. And since monotony is something to avoid up there, the possibility of using and "renewing" (through interchangeable wall panels) certain sections of the station (private sleep areas) is important. According to Dr. Yvonne A. Clearwater, NASA hopes to desing larger "private quarters" with "more amenities and a feeling of auditory and visual privacy". Such private areas are safety valves for coping with stress, and necessary retreating from one another must occur in the best personal "atmosphere" for each crew member, even by "creating" it, if necessary.

Constant activity is of course a good weapon against boredom, and as Soviet consmonaut Valeriy Ryumin said, "work is the best cure for anxiety and depression". But work rhythms can cause stress all the same, and therefore a good system to maintain both individual and group stability is surely that of alternating tasks and roles aboard, as far as possible. Of course hierarchy is important, but the possibility for each crew member to be safely and easily replaced by his colleagues is equally important for the organizational integration of the group. The group as such should be able even to replace a dropout, since deviance cannot be excluded a priori.

The most serious social stressor in space is and will be the prolonged separation from family, friends and society at large. Its best antidote is a wide spectrum of opportunities as well as protected places for frequent communication with loved ones on Earth. Also games to improve reflexes and concentration or just to relax are good antistressors, while we feel the use of drugs (the Russians used an extract of a plant related to gin-seng to increase their long-term stamina and resistance to stress and depression) might be dangerous, like every kind of doping.

Aboard a space station food and meals take on a psychological and social significance far beyond nutrition and caloric needs. Variety, change and self-selection as far as possible are important components for eating well. We must remember the highly positive value of this biological function for the individual as well as for the group. Eating together what we like means pleasure, relaxation, desire to communicate whith the others. Exactly what we need up there. And a "cultural diversity" in gastronomy for international crews may be stimulating for everybody aboard.

Both the United States and the Soviet Union have carried aloft passengers from other Countries, experiencing multi-national integrated activities aboard without problems. And last but not least, the Soviet cosmonauts became recently the protagonists of a very unusual experiment: testing yoga relaxation techniques in space with their Indian colleague as a trainer. Is individual and group stability obtainable in space through meditation and philosophy? Sure this could be possible, after all.

In conclusion, men are always the same here on Earth like up there in outer space, and our team's psycho-social approach and research, as well as the studies the American and the Russian space scientists are developing in this field, show that the farther we go from our planet and the longer we plan to stay away from it, the more we will have to know about ourselves. We may only hope that also ESA will face for Europe this new challenge.

References :

[1] Bedini, D., Space Station Hability Study: the Revelation between Volumes, Shapes & Colours and Human Behavior, IAF-IAA, 1986, 86-403

[2] Bluth, B.J., Human Systems Interfaces for Space Stations, AIAA/NASA, 84-1115

[3] Boeing Aerospace Co., National Behavior System - Soviet Space Stations A.S. Analogs/D180-28182-1, Oct. 1983

[4] Boeing Aerospace Co., National Behavior System - Space Station/Nuclear Submarine Analogs/CC0271, Oct. 1983

[5] Clearwarer, Y.A., Space Station Habitability Research IAF-IAA, 1986, 86-397

[6] Cohen, M.M., Space Station Program Implications for Human Factors

[7] Cooper, H.S.F. Jr., A House in Space New-York, 1976, Research/NASA 86702

[8] Gerber, J., Preliminary Analysis of Human Factors.Study of an analogous Environment: the Subamerines, ESA D/SSP-LTPO (86)1

[9] Goeters, K.M., The Recruitment and Organizational Integration of Space Personnel, IAF-IAA, 86-401

[10] Gunderson, E.K.E., Ed., Human Adaptability to Antarctic Conditions, Antartic Research Series, Vol. 22, Washington, D.C., American Geophysical Union

[11] Hall, H.T., The Hidden Dimension, New York, 1966

[12] Maruyama, M., Cultural Factors in High-Density Habitats, in Psyche and Design, pp. 11-32, W. Preiser, Ed., Orangeburg, N.Y., ASMER, Inc.

[13] Mayer Holzapfel, M., Abnormal Behavior in Zoo Animas, Philadelphia, 1968

[14] Natani, K. & Shurley, J.T., Sociopsychological Aspects of a Winter Vigilant South Pole Station, in Human Adaptability to Antarctic Condtions, Vol. 32, E.K. Gunderson, ed., Washington, D.C., American Geophysical Union

[15] Rasmussen, J., Ed., Man in Isolation and Confinement, Chicago, 1973

[16] Sanna, E., Cet animal est fou, Paris, 1976 (Italian Transl.: Lo Zoo Folle, Milano, 1977)

[17] Sells, S.B., The Taxonomy of Man in Eclosed Space, in Man in Isolation and Confinement, pp. 281-304, J. Rasmussen, Ed., Chicago 1973

[18] Stokols, D., On the Distinction between Density and Crowding: some Implications for Future Research, in Psychological Review 79, May, 274-277

[19] Watson, M.O., Proxemic Behavior/A Cross-Cultural Study, The Hague, 1970 (Italian transl.: Comportamento Prossemico, Milano, 1972)

[20] Wolf, M., Sociologie della Vita Quotidiana, Roma, 1979

Interventions

Jeff Hoffman

Bravo! I hope you and your colleagues can play an important role in the design of future space habitats. What is the most important thing in my opinion for human well-being in space is the ability to exercise creative activity. When human beings are treated as machines, they are unhappy and should be replaced by machines. In the long run, if we do not want creative minds in space, we should not send human beings.

Wubbo J. Ockels

- Relaxation is possible and very nice in weightlessness, but for sleep a confined feeling is of great help. I designed a new sleeping bag, it flew on D-1. USSR showed interest to fly it on MIR.

- Creativity is important! Spacelab is a good concept to allow significant work of man. Spacelab allows a strong interaction between astronauts and ground based scientists. It is the team of astronauts and scientists which guarantees the success of the investigations.

As the beauty of Space flight, especially the visual appearance of Earth, has already been presented in a splendid manner by my colleague Jeff Hoffman, NASA Astronaut, I would like to concentrate on a particular aspect of Space flight, the most impressive aspect, and, also the most difficult one to transfer. That is weightlessness.

Before I address this, I would like to justify astronauts who express the beauty of Space because this could be considered unfair. If there will be only a few who get the opportunity to go intospace, it is not nice to keep on stressing how wonderful it would be for the audience to experience Space. I am, however, convinced that not too far in the future, many will have the opportunity to buy a ticket for a space flight. To justify my optimism, please remember Lindberg who crossed the ocean for the first time by air in 1927. He flew from New York to Paris. Upon arrival a journalist asked him whether he thought normal people could fly across the ocean in the future. Lindberg answered: "I don't know. It will be too expensive and too uncomfortable". Within 50 years most people of the Western World could afford that flight, flying in a most comfortable jet airplane, now costing less than $ 1000. If Lindberg would have calculated, as a physicist, what amount of energy is needed to make that flight, using an ideal machine, the result would be only several hundred litres of gasoline. That would have indicated to him at what price eventually the man din the street could fly. It is just a matter of having the right machine.

The energy increase by going into space is only one half of the energy corresponding to Lindberg's flight. Of course we do not have the right machine now, the Shuttle is terribly inefficient. It takes millions of litres now. But I am convinced that not too far in the future a typical price to go to Space would be also about $ 1000. I therefore consider it absolutely fair to make you jealous about an experience of Space.

I continue here talking about gravity and the question which was the title of my previous presentation: "Is life earth-life"?

With life I mean the concept of life, the perception of changes of evolution, the expectation, the progress, all related with time. Is this concept general, or can it be different outside Earth? Do spaceflights cause us to think different about life? In my case it did. I was striken by the experience of gravity and weightlessness. The ideas, caused by this experience do not change my day to day life. I am not a philosopher, so please see my expressions as results of impressions of my flight and use it as inputs for <u>your</u> thoughts.

My impressions do not come only from the Space flight itself. They also come from others: the preparation of a Space flight causes already many changes in day to day life, the amount of travel, in my case I crossed the ocean about 50 times before my Space flight, the publicity, so many people place you on a special platform and push you up, to then pull on your attention from all sides.

The first time thoughts were triggered related to the difference in perception of forces versus acceleration was during one of the tests of the astronaut selection process. The test was riding a centrifuge providing 3 g's during 20 minutes. I was sitting in a dark enclosed capsule. In itself the experience was not unpleasant. I remember that I was pleased to feel the tumbling, the Coriolis effect on my vestibular organ when turning my head, although that feeling itself is not pleasant. But suppose I would have accelerated linearly! 3-g's after 20 minutes would have given me an enormous speed! It was comforting to notice that I was turning around in circles.

The ideas developed during the preparation of the flight got confirmed during my Space flight. I present these ideas not as a physicist. They are not completely logical. They would be destroyed if I would require consistency with all the physical knowledge I have. I do not want to kill a thought or impression by saying "nonsense". I _have_ experienced the difference between free fall and gravity force and that _did_ cause me to raise the question "Is life Earth-like"?. That the man-kind is Earth-like that is obvious. That our legs are about 1 m. long, that we need the amount of muscles to fight gravity. That the beat of music is about 0.5 sec. is related to our body size and the force gravity. At the Moon the natural beat would be much longer, the music therefore much slower. Gravity brings speed, is hectic. I believe in harmony of life and its environment. Many aspects of life when seen, will appear to be related via a natural and logical sense to its environment. Even the human communication capabilities. We are sensitive to delays in our communication, think about a telephone conversation with a good friend across the ocean. Especially emotional contact will be very difficult because of the delay. Is it a coincidence that the delay to which we are sensitive, e.g. about 0.1 sec., corresponds with a distance of the typical Earth dimension?

I will not, however, address in what sense the biological appearance of us is Earth-like. No, I want to address the question whether our concept of life is Earth-like.

To share my reflections on this question I will read to you my piece of the book from the Association for Space Explorers, a book made

Arnold H.W. Woodruff

I was very interested to have your comments on the psychological effects of working in isolated environement -on nuclear submarines, Antartic bases.
As a scientisit, a geophysicist, I spent 2 seasons in the Antarctic: up to 3 months at a time in the field with 1 or 3 other people. The nuature of our turning required us to maintain 24 h surveillance on, at least, the whereabouts of our 1 companion in each 2 men unit. This was a safety requirement of course vis-a-vis crevasses, etc.. Nevertheless on returning to Base, or ship one of the first things many of us wanted to do was to tuck oneself away in a root or cranny, perhaps with one's personal mail, a pipe and one's thoughts -to relax by oneself. This was a refreshing environment and necessary experience.

I would certainly therefore, from my own experience, like to support your suggestion that lack of a private individual retreat may cause psychological probles if closely knit communities are isolated through space travel for long periods.

The cost may be lightly greated but in the long term I think a reasonable expense. Particularly so perhaps as most future space travellers as current day Astronauts, and Antartic Explorationists and latter day ocean travellers tend to draw their numbers for those of a certain pioneering explorationist mentality -people who wish to pit their wits and abilities not only in the quest for knowledge but also agains nature and the universe, with which at times they would probably like, perhaps more than most, to be alone.

Henri Augustin

En tant que météorologiste, je me préoccupe de l'atmosphère dans votre station spatiale (qualité de l'air et climat).

- la présence de plantes va y introduire des pollen auxquels "l'homo cosmicus" risque d'être allergique. Il faudrait examiner les moyens de les éliminter, avec les autres poussières et pollutions possibles ;

- le climat va-t-il être absolument uniforme ? Si tel était le cas, il serait intéressant de le moudler en température, humidité et vent pour éviter la monotonie.

CLOSING SPEECH

Hubert Curien
Professor
Université de Paris VI
4, place Jussieu 75005 Paris, France

It is my very pleasant task to thank and congratulate the organisers of this Colloquium : the European Space Agency, through its Director General, Reimar Lüst, and Director of Science, Roger Bonnet, and the European University of Philosophy, through its Director, Jean-Pierre Faye. The linchpin in the preparations for this gathering, to whom we are greatly indebted, was Jean-Claude Pecker, assisted by Jean Schneider. Monique Orine and her team have shown admirable efficiency in coping with the practical difficulties of running this event. We owe thanks to NASA, represented by the astronaut J. Hoffman, courtesy of whom we have seen a display of quite remarkable images, and to the USSR Academy of Sciences, represented by Professor Ursool. Finally, the Ministère de la Recherche et des Enseignements Supérieurs has acted as host for our meetings and the Observatoire de Paris for our other business, and I speak for all of us in gratefully acknowledging their hospitality.

This meeting has been the outcome of a unique idea, the idea of bringing together astrophysicists, space scientists and engineers, and philosophers, and it has lived up to expectations, giving rise to a most fruitful exchange among the representatives of all these disciplines.

Rather than attempt to summarise the proceedings, invevitably failing to reflect their diversity and spontaneity, I prefer to confine myself to offering a few comments on those papers or discussions which, for me, left the most abiding impressions.

Several speakers gave demonstrations of the art of extrapolating without going too far. The paper by H. C. van de Hulst was an object lesson in this respect. Unstable or merely complex systems are now being attacked with great success, this as a result of the introduction of new mathematical concepts and methods. These selfsame concepts can themselves be put to the test by interpretation of phenomena other than those which govern the behaviour of inert matter. Living beings and social structures offer attractive scope for this kind of exercise.

This was of course a prime occasion for pondering on the merits of the various ways in which we study the world. The analytical method aims to probe at matter until it has yielded up all its secrets. Physicists for instance are engaged in dissecting what were hitherto

J. Schneider and M. Léger-Orine (eds.), Frontiers and Space Conquest, 247–248.
© 1988 by Kluwer Academic Publishers.

called elementary particles, in the hope of finding particles even more elementary. But approaches based on globalisation have a lot to offer as well. A broad-sweep view of the planets, especially such a view of Earth, with the details removed, brings out global phenomena <hich are most instructive in their own right. By blocking out a mass of fragmentary perceptions, the researcher's eye perceives original aggregate phenomena. In an approach combining analysis and globalisation the structural schemes devised by mathematicians and logicians become extremely useful. Is it reasonable to hope that the "information society" towards which humankind is evolving, as outlined for us by André Lebeau, will be better equipped to get things in proper perspective since it will be able to hand over the tasks of recording, classifying and processing matters of detail to the computer?

This brings me to the subject of relations between man and machine in this new kind of society, and here we come face to face with some important questions. The ever-widening information gap between laymen and specialists, to say nothing of the difficulties of communication between specialists in different fields, is a matter of concern for anyone with pretensions to take an intelligent interest in today's social panorama. The methods we use in education and training are in urgent need of review.

The conquest of space has been central to our discussions. As H. C. van de Hulst put it, where does Utopia start? One point is clear: a common logic informs all projects and programmes, wherever they be developed, in the USA, the URSS, Europe or elsewhere. The same stages have to be gone through. The same objectives are envisaged for the same first steps. These stages, of whose feasibility we can be virtually certain as of now, will take perhaps the next fifty years to get through. And after that? We have each of us offered our best guesses. They could not have been more varied!

Let us make a modest attempt to project ourselves forwards fifty years. The economic, scientify and intellectual arguments in support of the conquest of space reveal weaknesses under analysis. Over the long term, to be frank, there are no cast-iron economic arguments. The more sensible approach is to attempt to justify each stage individually, so that an optimum date can be set for its completion. And even this can only be done according to essentially relative criteria, in which the part played by the fortuitous, the emotive and irrational can be as decisive as it is unpredictable.

In offering these few thoughts I have definitely not been trying to sum up our discussions, still less to offer any conclusion. This gathering has had the very considerable merit of wresting us away from our usual day-to-day concerns and prompting us to pay attention to considerations on a somewhat higher plane.

INDEX
(English)

INDEX
(français)